Lecture Notes in Economics and Mathematical Systems 451

Springer
Berlin
Heidelberg
New York
Barcelona
Budapest
Hong Kong
London
Milan
Paris
Santa Clara
Singapore
Tokyo

Matthias Salge

Rational Bubbles

Theoretical Basis, Economic Relevance, and Empirical Evidence with a Special Emphasis on the German Stock Market

Springer

Author

Dr. Matthias Salge
Kirchhainer Strasse 44
D-60433 Frankfurt/Main
Germany

Cataloging-in-Publication Data applied for

Die Deutsche Bibliothek - CIP-Einheitsaufnahme

Salge, Matthias:
Rational bubbles : theoretical basis, economic relevance, and empirical evidence with a special emphasis on the German stock market / Matthias Salge. - Berlin ; Heidelberg ; New York ; Barcelona ; Budapest ; Hong Kong ; London ; Milan ; Paris ; Santa Clara ; Singapore ; Tokyo : Springer, 1997
(Lecture notes in economics and mathematical systems ; 451)
Zugl.: Kiel, Univ., Diss., 1996
ISBN 3-540-62629-8
NE: GT

ISSN 0075-8442
ISBN 3-540-62629-8 Springer-Verlag Berlin Heidelberg New York

Typesetting: Camera ready by author
SPIN: 10546757 42/3142-543210 - Printed on acid-free paper

Acknowledgements

This paper would not have been completed without external help.

I am gratefully indebted to my supervisor Prof. Dr. Seidl for his continuous support of my research activities during my years at the 'Institut für Finanzwissenschaft und Sozialpolitik'. Especially, from the discussion within various research seminars of this institute I have benefited immensely. I would like to thank Prof. Dr. Hansen for improving my insights in statistical and econometrical issues. To both researchers I owe thanks for valuably commenting and refereeing this thesis.

Furthermore, I want to thank Kathrin Fach for reading and correcting an earlier draft of this book. Thom Stevens deserves special thanks for improving my English style, without doubt a hard job.

Finally, this paper is dedicated to my wife Gabi for her patience and assistance while this paper was completed.

December, 1996 Matthias Salge

Contents

Chapter 1

Introduction

The evaluation of assets belongs, without doubt, to one of the most fascinating areas in economic theory and practice. This fascination may partly be fueled by the fact that many asset markets seemingly promise an opportunity for becoming immensely rich as many fabulous stories about investors would suggest[1]. This may also be an important reason for the attractiveness of asset pricing models. In particular, economic research has spent a lot of time on designing elaborate models which aim to enable an ex-post description as well as a profitable prediction of asset prices. However, many of the traditional models often fail to explain or forecast the movement of asset prices accurately. Those models were mainly based on a fundamental evaluation related to a stream of cash flows connected with the asset under consideration. This failure inspired the search for other factors, beyond fundamentals, which may possibly be responsible for the apparent deviation of market prices from their fundamentally justified prices.

Rational bubbles, which first appeared in the economic world at the end of the 1970s[2], represent one possibility for explaining the observed excessive

---

[1]On the other hand, besides this apparently existing opportunity for spectacular capital gains, sharp losses in individuals' wealth induced by extraordinary declines in asset prices have been observed in financial history confirming the view that the chance of obtaining higher returns is acquired by bearing a higher risk.

[2]The idea of bubbles is much older, however, the formal framework, as it is presented here, occurred within rational expectation models in this period.

fluctuations in asset prices. Rational bubbles can provisorily be defined as a speculative markup on asset prices generated by self-sustaining expectations driving up prices without any further fundamental justification[3]. Explaining the apparent excess volatility on asset markets represents one reason why the notion of bubbles has become popular in economic theory. Bubbles are additionally able to capture the phenomenon of risk increasing crashes, i.e. a sharp decline in prices usually connected with extreme losses in wealth. This close relationship between the phenomenon of crashes and the notion of bubbles can also be confirmed by considering the history of famous bubbles. The first historical episode, which is usually ascribed to the existence of bubbles, explicitly describes a crash-like phenomenon, the tulipmania in the Netherlands between 1634 and 1637 when trading new tulip varieties previously imported from Turkey in mid-1500s became extremely fashionable. In particular, rare tulip bulbs were sold at high prices apparently without any further fundamental justification[4]. In 1636 speculators, attracted by the extreme price increase, entered the tulip market inducing further price upswings. This effect may well be captured by the concept of bubbles stating that self-fulfilling movements away from the fundamentals may be reasonable when prices are expected to increase any further. However, in February 1637 prices suddenly broke down and tulip bulbs could not be sold at 10 % of their peak prices. In subsequent years the price decline continued and caused a deep recession in the Netherlands[5].

---

[3]See Garber (1990b), for a similar definition. Based on this preliminary definition, the main problem connected with bubbles becomes already obvious. The notion of bubbles characterized as the difference between actual market price and fundamental price is a relative concept depending on the correct characterization of the underlying fundamentals. One may falsely attribute excess volatility to the existence of bubbles while only fundamentals are incorrectly defined. This issue will be discussed more extensively all over the book here.

[4]However, this view has been disputed recently. Rumors concerning increased demand in France induced by woman, who attached tulips to their gowns, may serve as a fundamental reason for the observed price behaviour in this period, as Camerer (1989) pointed out. Garber also doubts the lack of a fundamental explanation and gives some evidence for his conviction in his (1990a) paper.

[5]Other episodes historically associated with bubbles involve the Mississippi Bubble (1719-20) and the South Sea Bubble (1720). See Garber (1990a) for a description of both phenomena. Furthermore, a spectacular price increase for yearlings thoroughbreds

Other famous events usually associated with the presence of bubbles are more closely related to capital markets, i.e. extreme break downs in stock prices. In particular, the crash in October 1929 when stocks lost more than a third of their previous value[6] and the 1987 crash when the highest one–day price drop of 22.6 % was observed on October, 19, have often been explained with the existence of bubbles[7]. However, it will turn out in the following analysis that bubbles are principally able to capture effects beyond crashes observed in stock prices.

In general, two lines of research concerning bubbles can be distinguished: A first approach examines the effects of bubbles in a general equilibrium setting. Based mainly on models with overlapping generations, researchers aim to answer questions like whether bubbles can actually exist, and if so, how they influence the economy as a whole. In particular, their impact on the capital stock, i.e. the long run growth of the economy is focussed on explicitly. Tirole (1982, 1985), Obstfeld, Rogoff (1983, 1986) and Weil (1987, 1989) among others follow these lines by examining the real effects of bubbles in general equilibrium models. We partly present their results in section 2.4.4. However, the analysis here mainly relies on the second approach which concentrates on the possibility of bubbles in partial markets and tries to examine their existence empirically. This line may be justified by the conviction that investigating the general equilibrium implications of bubbles appears to be unreasonable if the existence of bubbles cannot be proven empirically.

The intention of the analysis here is threefold: First, we aim to broaden the set of bubbles discussed in literature so far. Therefore, chapter 2 provides the theoretical framework for deriving bubbles from a simple stochastic difference equation. A general procedure is presented which enables us to deduce bubble processes systematically. It will turn out that by ap-

---

in 1970s attributed to the existence of bubbles is reported by Camerer (1989).

[6]See, for instance, De Long, Schleifer (1991) and Rappoport, White (1994) for an examination of the 1929 crash.

[7]See, for instance, White (1990) who provides some evidence in favor for the presence of a bubble in the 1929 crash and Shiller (1989b) who, based on survey data, ascribes the 1987 crash to phenomena beyond a fundamental explanation. For a further investigation of the 1987 crash see Schwert (1990a, 1990b).

plying this procedure, all bubble specifications analyzed in literature until the present are obtained. Simultaneously, the set of bubble processes is extended significantly. Second, the economic relevance of bubbles will be clarified in chapter 3, discussing the three most popular applications of bubbles in economic theory, i.e. bubbles on capital markets, on foreign exchange markets and in hyperinflationary periods. In particular, in the context of capital markets, a so–called 'bubble cycle' is provided which describes the development of bubbles in periods of crashes more precisely. Third, possible testing strategies for gaining evidence of the existence of bubbles are described in chapter 4 while chapter 5 investigates the evidence of several bubble specifications on the German Stock Market.

Chapter 2

On the Theoretical Derivability of Rational Bubbles

Consider as a starting point for our analysis of rational bubbles the following simple stochastic linear difference equation (2.1) where the current value of the variable y_t assumed to be endogenous depends on the expectation of that endogenous variable one period ahead and an exogenous variable[1].

$$y_t = \alpha y^e_{t,t+1} + x_t \tag{2.1}$$

The expression $y^e_{t,t+1}$ in equation (2.1) represents the subjective expectation of the endogenous variable y formed at time t for the next period, x_t denotes a possibly random exogenous variable which may additionally include a random component while $\alpha \neq 0$ is a given parameter.

While we will justify equation (2.1) by economic grounds in chapter 3, a somewhat philosophical intuition for considering equation (2.1) may be given here: Taking a look at the current state of the world, we can ask the

[1]Notice that the distinction between endogenous and exogenous variables inserts a well known causality inside the model which is normally based on economic reasons: the endogenous variable represents the variable to be determined by the exogenous variable whereas the exogenous variable is determined outside the model. This becomes relevant when testing for rational bubbles in chapter 5.

subtle question by which factors the present can generally be influenced. Doubtlessly, the past determines to some extent the current state of the world. This is mainly considered in econometric time series analysis where the time path of a variable is explained by its past realizations (e.g. variables represented by an ARIMA–process which plays a prominent role in econometrics). In equation (2.1) the impact of the past is captured by the exogenous variable. Further, it is assumed that the information contained in prior realizations of the exogenous variable, which is viewed as the driving variable of the model, is already incorporated in the current value of the exogenous process x_t to most extent[2]. Besides the past, one must also admit that the present is likely to be affected by the future, as well. All intertemporal decision processes lead individuals not only to include past realizations of an objective in question but to take expectations regarding future realizations of that objective. Since the future is inherently unknown, subjective expectations concerning future states of the world become relevant. Therefore, it can be reasonable to include expectations of a variable considered in a model which tries to explain the present. Since taking into account future values of the corresponding endogenous variable is more interesting in comparison to future values of the exogenous variable, which is fully determined outside the model, equation (2.1) integrates the expected value of the endogenous variable y_t one period ahead.

Solving equation (2.1) means characterizing the endogenous variable y_t in terms of the exogenous variable x_t and the parameter α simultaneously satisfying equation (2.1). In order to make equation (2.1) solvable, we have to determine the way subjective expectations are formed. Various expectation formation processes are suggested in literature. They have in common that they all are based to some extent on the past. But they differ in the way the past enters the expectations formation process. The assumption of adaptive or extrapolative expectations, for instance, at first frequently used in economic theory explains expected values of a variable through a weighted average of its own past realizations. The resulting geometric lag representation of adaptive or extrapolative expectation is due to a revision process which updates the change in the expectations

[2]This corresponds to the martingale concept introduced later where the past is entirely incorporated in the current value of that process.

formation only by a percentage less than a hundred of the previous forecast error[3]. Consequently, a systematic failure in forecasting remains. This undesirable property of adaptive or extrapolative expectations motivated the development of the rational expectations hypothesis. Since the notion of rational expectations refers to conditional expectation of statistic theory, some comments on the underlying statistics will be useful here[4].

2.1 Statistical framework

Based on an axiomatic approach to probability, consider a probability space $(\Omega, \mathcal{F}, P)$, where Ω denotes a sample space (i.e. the set of all possible outcomes ω), $\mathcal{F}$ (assumed to be a σ-algebra[5]) represents a collection of all events of interest (each a subset of Ω) and P stands for a probability (set) function mapping all elements of $\mathcal{F}$ into the interval $[0,1]$ and satisfying the Kolmogorov axioms.

A discrete random variable X defined on the probability space $(\Omega, \mathcal{F}, P)$ is a (set) function which associates each element ω of the sample space Ω with a real number

$$X : \Omega \to \Re^1 \mid \omega \in \Omega \to X(\omega) \in \Re^1$$

such that the set $X^{-1}(R_X) = \{\omega \mid X(\omega) \in R_X\}$ is contained in $\mathcal{F}$ for all

[3]Adaptive expectations, for instance, can be characterized as $y^e_{t,t+1} - y^e_{t-1,t} = \phi(y_t - y^e_{t-1,t})$ with $0 < \phi < 1$. Notice that the term in parenthesis on the right side represents the forecast error and ϕ the percentage of revising expectations. Simple arrangement of the above equation yields $y^e_{t,t+1} = \phi y_t - (1-\phi) y^e_{t-1,t}$. Hence, the subjective expectation for $t+1$ is a weighted average between the actual value y_t and the expectation for period t formed at period $t-1$. Calculating the forecast error $\epsilon_{t+1} = y_{t+1} - y^e_{t,t+1}$ and substituting the definitorial equation for adaptive expectations successively we obtain a geometric lag structure $\epsilon_{t+1} = y_{t+1} + \beta_1 y_t + \beta_2 y_{t-1} + \ldots$ where $\beta_i < 1 \; \forall i \geq 1$ are functions of ϕ. Concluding, by using adaptive expectations, the forecast error contains systematic and, therefore, predictable components.

[4]For a more advanced discussion of the following statistic topics see Spanos (1986) or any other textbook of statistics.

[5]A set $\mathcal{V}$ is said to be a σ-algebra with respect to Ω if it satisfies the following properties: (i) $\Omega \in \mathcal{V}$, (ii) $V \in \mathcal{V}$ with $V \subset \Omega \Rightarrow \bar{V} \in \mathcal{V}$, (iii) $V_1, V_2, \ldots, V_n, \ldots \in \mathcal{V}$ with $V_1 \subset \Omega, V_2 \subset \Omega, \ldots, V_n \subset \Omega, \ldots \Rightarrow \cup_{i=1}^{\infty} V_i \in \mathcal{V}$

subsets R_X of the real line[6]. The last condition guarantees the preservation of the event structure after relating the sample space Ω and the random variable X.

In the following we introduce some expectational concepts of discrete random variables on which the analysis in the remaining chapters is mainly based[7]. Provided that all elements of the sample space Ω are contained in the σ–algebra $\mathcal{F}$[8], the *(unconditional) mathematical expectation* of a discrete random variable X is given by

$$E[X] = \sum_{\omega \in \Omega} X(\omega)P(\omega)$$

The (unconditional) expectation $E[X]$ can be interpreted as the average of the values taken by the random variable X evaluated over the whole sample space. Each element of Ω represented by the random variable is weighted by its probability. The concept of expectation can easily be extended to subsets of the sample space Ω. Let A be a subset over Ω with $A \subset \mathcal{F}$ and $P(A) \neq 0$. The expectation of a discrete random variable X related to the subset A of the sample space Ω is given by

$$E[X \mid A] = \frac{1}{P(A)} \sum_{\omega \in A} X(\omega)P(\omega)$$

The expectation $E[X \mid A]$ is called *conditional expectation* with respect to subset A. The conditional expectation of a random variable equals the mean value of the random variable after occurrence of event A: only those realizations of X are taken into account which are compatible with the subset $A \subset \Omega$.

So far, we have considered only one random variable X and have specified the unconditional expectation and the expectation conditional on the realization of a subset A of the sample space Ω. Assume now a vector of n

---

[6]These subsets are called Borel fields. A Borel field is formally defined by $R_X = \{r \mid r \leq X\} \quad X, r \in \Re^1$. Further, the set $\sigma(X) = \{X^{-1}(R_X) \mid \forall R_X\} \subset \mathcal{F}$ is called the smallest σ–algebra generated by X. It can be thought of the portion of information obtainable from the knowledge of X.

[7]For a more general approach including both discrete and continuous random variables see any textbook of statistics, for instance Spanos (1986). Broze, Szafarz (1991) p 2 discuss the relevance of these concepts in the context of rational expectations models.

[8]This is usually assumed to be satisfied since otherwise no probability $P(\omega)$ would exist.

random variables $(X_1, X_2, ..., X_n)$ defined over the same probability space $(\Omega, \mathcal{F}, P)$. The *(unconditional) mathematical expectation* of a discrete random vector $(X_1, X_2, ..., X_n)$ provided that all elements of the sample space Ω are contained in the σ–algebra $\mathcal{F}$ is given by

$$E[X_1 X_2 \dots X_n] = \sum_{\omega \in \Omega} X_1(\omega) X_2(\omega) \dots X_n(\omega) P(\omega)$$

Analogously, the (unconditional) expectation $E[X_1 X_2 \dots X_n]$ can be interpreted as the average of the values of the random vector $(X_1, X_2, \dots, X_n)$ evaluated over the whole sample space.

If we are interested in the expectation of X_{n+1} conditional on the special subset $(X_1, X_2, ..., X_n)$ of the above collection of random variables[9], the nature of A must be characterized more precisely in the multivariate case. Knowing the precise realizations of the random variables $(X_1, X_2, ..., X_n)$ denoted by $(x_1, x_2, ..., x_n)$, let A be a subset over Ω with $A \subset \mathcal{F}$ corresponding to all outcomes ω which are consistent with these realizations[10]. Formally, we can define A as a function of $(x_1, x_2, ..., x_n)$ as

$$A(x_1, x_2, ..., x_n) = \{\omega \mid X_1(\omega) = x_1, X_2(\omega) = x_2, ..., X_n(\omega) = x_n\}$$

For $P(A) \neq 0$ the *conditional expectation* of a discrete random variable X_{n+1} related to the subset A of the sample space Ω is given by

$$E[X_{n+1} \mid A(x_1, x_2 ..., x_n)] = \frac{\sum_{\omega \in A(x_1, x_2 ..., x_n)} X_{n+1}(\omega) P(\omega)}{P(A(x_1, x_2 ..., x_n))}$$

Since A depends only on $(x_1, x_2, ..., x_n)$, we can rewrite conditional expectation as

$$E[X_{n+1} \mid X_1(\omega) = x_1, ..., X_n(\omega) = x_n] = E[X_{n+1} \mid x_1, \dots, x_n]$$

which is a more familiar notation. Conditional expectation gives us the mean value of X_{n+1} after $(x_1, x_2, ..., x_n)$ has been observed: Calculation

[9]The following definition of conditional expectation can easily be extended to more general subsets of the above collection of random variables, but the following definition will be sufficient in the analysis of time paths of random variables in the analysis below.

[10]This characterization is due to Broze, Szafarz (1991) p 3.

of conditional expectation includes only those possible outcomes of X_{n+1} which are compatible with the observation of $(x_1, x_2, ..., x_n)$[11].

So far, we have considered a probabilistic framework where the vector $(X_1, X_2, \ldots, X_n)$ represents different (static) characterizations of a phenomenon in question. This concept can easily be translated into a dynamic context by reinterpreting the random vector $(X_1, X_2, \ldots, X_n)$ as standing not for different but for one and the same characteristic X at different points of time $(X_{t_1}, X_{t_2}, \ldots, X_{t_n})$[12]. Thus, the above considered random vector $(X_1, X_2, \ldots, X_n)$ can be interpreted as what is called a stochastic process consisting of a sequence of time–dependent random variables.

Definition 2.1 A *stochastic process* is a sequence of random variables $\{X_t, t \in \mathcal{T}\}$ defined on the same probability space $(\Omega, \mathcal{F}, P)$ where t denotes a time subscript and $\mathcal{T}$ a time horizon[13].

Under certain consistency conditions[14] the (joint) distribution function of a finite number of random variables $(X_{t_1}, X_{t_2}, \ldots, X_{t_n})$ exists and characterizes completely the nature of the stochastic process $\{X_t, t \in \mathcal{T}\}$, although the sequence of random variables $\{X_t, t \in \mathcal{T}\}$ may refer to an infinite time horizon $\mathcal{T}$. This property enables us to define all moments of any subset

[11]Notice that if we assume X_{n+1} to be independent of $(X_1, X_2, ..., X_n)$, the conditional expectation $E[X_{n+1} \mid A(x_1, x_2, ..., x_n)]$ is equivalent to the unconditional expectation $E[X_{n+1}]$. Assuming dependence of X_{n+1} from $(X_1, X_2, ..., X_n)$, the conditional expectation differs from the unconditional expectation and approaches the unconditional expectation $E[X_{n+1} \mid A(x_1, x_2, ..., x_n)] \rightarrow E[X_{n+1}]$ if A approaches the sample space $(A \rightarrow \Omega)$.

[12]Notice that a corresponding reinterpretation of the sample space would redefine Ω as the set of all possible states of the world.

[13]Notice that for $\mathcal{T}$ being countable (e.g. $\mathcal{T} \subseteq Z$) the stochastic process is a discrete process. If $\mathcal{T}$ is uncountable (e.g. $T = [a; b]$ for $a, b \in \Re$), the stochastic process is called continuous.

[14]According to Kolmogorov, these are the symmetry condition and the compatibility condition. The symmetry condition requires that the distribution of the stochastic process $\{X_t, t \in \mathcal{T}\}$ remains unchanged if any permutation of the indices of the random variables takes place. The compatibility condition demands the possibility of reducing the dimension of the joint distribution of the random variables through marginalisation. See Spanos (1986) p 133 or Malliaris and Brock (1982) p 33.

$(X_{t_1}, X_{t_2}, \ldots, X_{t_n})$ of the stochastic process $\{X_t, t \in \mathcal{T}\}$ and the analysis of conditional expectation applies.

The notion of a stochastic process $\{X_t, t \in \mathcal{T}\}$ can easily be extended to the multivariate case if we take into account more than one stochastic process. For a *vector stochastic process* $\{\mathbf{X_t}, t \in \mathcal{T}\}$ with $\mathbf{X_t} = (X_{1t}, X_{2t}, \ldots, X_{nt})$ where X_{it} is a single stochastic process we can derive analogous expressions for distributional moments and conditional expectations. For instance, consider two stochastic processes $\{X_t, t \in \mathcal{T}\}$ and $\{Y_t, t \in \mathcal{T}\}$. Suppose we are interested in the expectation of Y_{t+1} conditional on past realizations of its own and on past realization of the somehow connected stochastic process $\{X_t, t \in \mathcal{T}\}$. Denoting these known prior realizations by small letters we can characterize the subset A over Ω with $A \subset \mathcal{F}$ as

$$\begin{aligned} A(y_t, y_{t-1}, \ldots, x_t, x_{t-1}, \ldots) = \{\omega \mid\ & Y_t(\omega) = y_t, Y_{t-1}(\omega) = y_{t-1}, \ldots, \\ & X_t(\omega) = x_t, X_{t-1}(\omega) = x_{t-1}, \ldots\} \end{aligned}$$

Then, the *conditional expectation* of a discrete random variable Y_{t+1} related to the subset A of the sample space Ω is given by

$$\begin{aligned} (2.2) \qquad & E[Y_{t+1} \mid A(y_t, y_{t-1}, \ldots, x_t, x_{t-1}, \ldots)] \\ & = \frac{\sum_{\omega \in A(y_t, y_{t-1}, \ldots, x_t, x_{t-1}, \ldots)} Y_{n+1}(\omega) P(\omega)}{P(A(y_t, y_{t-1}, \ldots, x_t, x_{t-1}, \ldots))} \end{aligned}$$

This expression is useful in the analysis of bubbles in the following analysis[15].

[15]For a more detailed treatment of vector stochastic processes see Spanos (1986) p 135. Notice that the main problem in deriving the explicit distribution of each X_t at any point of time t, is that only one observation of the random variable X_t at time t is available. Consequently, restrictions on the stochastic process $\{X_t, t \in \mathcal{T}\}$ must be imposed in order to be able to determine the distribution of X_t at time t or, at least, their first moments. Usually restrictions concerning the time-homogenity and the memory of a stochastic process $\{X_t, t \subset \mathcal{T}\}$ are suggested. See Spanos (1986) p 137.

2.2 Rational expectations

As already mentioned above, the concept of rational expectations[16] is related to the concept of conditional expectation of statistic theory. In order to describe rational expectations rigorously, recall the definition of conditional expectation in equation (2.2) in terms of two stochastic processes $\{y_t, t \in \mathcal{T}\}$ and $\{x_t, t \in \mathcal{T}\}$[17]. In economic theory the set A containing past realizations of both the exogenous variable x and the endogenous variable y up to time t is called the set of available information[18] at time t denoted by I_t. Principally, two situations concerning the composition of information set I_t corresponding to our rational expectations model are distinguished in literature[19]:

- I_t consists of all current and past values of both the exogenous variable x and the endogenous variable y from the beginning of an observation period $t = 0$ up to present time t. Formally, for $\forall t$

  $$I_t = \{x_t, x_{t-1}, \ldots, x_0, y_t, y_{t-1}, \ldots, y_0\}$$

  In this case, individuals are assumed to have 'no loss of memory'. The entire time path of all relevant variables is completely known. Since only the exogenous variable x is assumed to be relevant for forecasting the endogenous variable y, this is actually the largest possible information set consisting of variables included in the model and definitely describes the best information situation.

---

[16]For a more extended presentation of the notion of rational expectations see Begg (1982) or Sheffrin (1983).

[17]In the following it will not be distinguished between random variables previously denoted by capital letters and their realizations denoted by small letters. Both random variables and their realizations are now written in small letters bearing in mind that all future values refer to random variables and past and present values refer to realizations of random variables. The present is always related to time t.

[18]Notice that current values of both the exogenous variable x and the endogenous variable y are incorporated in the information set I_t. Depending on the time structure of the model which determines the knowledge of the variables at a certain point of time, current values can also be excluded.

[19]For a more general description of the information set in the case of vector stochastic processes see Broze, Szafarz (1991) p 5.

- Alternatively, I_t may contain all current and past values of both the exogenous variable x and the endogenous variable y, but only up to a certain time lag k, $t > k > 0$. Formally, for $\forall t$

$$I_t = \{x_t, x_{t-1}, \ldots, x_{t-k}, y_t, y_{t-1}, \ldots, y_{t-k}\}$$

This concretion of I_t describes a 'loss of memory'–situation. All realizations of both the exogenous variable x and the endogenous variable y beyond period $t-k-1$ are not available. Notice that the length of the period $t-k$, up to which realizations of the variables are considered, remains equal when times go by[20].

An important premise for dealing with rational expectations models is that information sets are growing with time. This is due to the fact that many interesting properties of conditional expectations require increasing information sets. Notice that only the first characterization of an informational situation above allows for increasing information sets ($I_t \subset I_{t+1}$). That is mainly the reason why we will assume this informational situation in the following analysis of bubbles[21].

After characterizing the information set I_t as an increasing one, we are now able to define rational expectations more precisely.

Definition 2.2 Let $\{I_t, t \in \mathcal{T}\}$ denote an increasing sequence of information sets which consist of past realizations of all variables of a model under consideration. We define *rational expectations* of a variable of that model formed by an individual at time t as the

[20] The impact of every single process on the memory of the system represented by the parameter k can differ between both variables. Here, we consider k to be equal for both processes on simplicity grounds.

[21] The first characterization of I_t may serve as a reference point for all I_t considered later. For instance, I_t above can be extended by the inclusion of sunspots. Sunspots can be viewed as variables outside the model from which people think that they have the power to influence the endogenous variable. For a more detailed treatment of sunspots see section 2.4.3. Moreover, an even smaller information set than that defined here called the *minimal information set* may also be sufficient if past values of the endogenous variable y can fully be determined by using past values of the exogenous variable x.

mathematical expectation conditional on the information set I_t[22], i.e.

$$\underbrace{y^e_{t,t+1}}_{\text{subjective expectation}} = \underbrace{E[y_{t+1} \mid I_t]}_{\text{rational expectation}}$$

In our case, for instance, the rationally expected value of the endogenous variable y one period ahead is the mathematical expectation conditional on past and current realizations of both the exogenous variable x and the endogenous variable y contained in the information set I_t.

$$E[y_{t+1} \mid I_t] = E[y_{t+1} \mid y_t, y_{t-1}, \ldots, y_0, x_t, x_{t-1}, \ldots, x_0]$$

Notice that unlike other expectation formation processes, rational expectations depend only indirectly on the past through the information set I_t.

The notion of rational expectations is due to Muth (1961). In his seminal paper he emphasizes the importance of dynamics for economic theory. Taking into account future variables in economic modelling, expectations become relevant. Since expectations themselves are unobservable, the way of forming expectations has to be defined. Among other concepts of expectation formation processes[23] Muth (1961) suggests the rational expectations hypothesis: individuals form their expectations rationally which means that their subjective expectation equals conditional expectation.

The characteristics of the rational expectations hypothesis can best be described in terms of the rational prediction error. The expression 'prediction' indicates that rational expectations of a variable can be thought of as representing a forecast since in statistic theory forecasts made by using rational expectations coincide with the forecasts based on conditional expectation.

Definition 2.3 Consider a stochastic process $\{y_t, t \in \mathcal{T}\}$ and an increasing sequence of information sets $\{I_t, t \in \mathcal{T}\}$. A *rational pre-*

[22] A more general approach assumes the subjective probability distribution to be the same as the conditional distribution based on information set I_t, see for instance, Lucas, Prescott (1971). For our purpose the consideration of first moments is sufficient.

[23] Besides the notion of rational expectations, Muth (1961) examines also the above mentioned adaptive or extrapolative expectations.

diction error ϵ_{t+1} is the difference between the actual realization of that process at $t+1$ and the rationally expected value. Formally, for $\forall t$

$$\epsilon_{t+1} = y_{t+1} - E[y_{t+1} \mid I_t] \tag{2.3}$$

where $E[y_{t+1} \mid I_t]$ denotes the conditional expectation of the random variable based on information set I_t.

Notice that a rational prediction error represents a subcase of the more general concept of a forecast error defined as $y_{t+1} - \hat{y}_{t,t+1}$ where $\hat{y}_{t,t+1}$ stands for an arbitrary forecast of y_{t+1} at t. In the case of a rational prediction error $\hat{y}_{t,t+1}$ is replaced by $E[y_{t+1} \mid I_t]$.

Before presenting two interesting properties of a rational prediction error, we must state a proposition which will prove very useful in the analysis of rational expectations models. As mentioned earlier, we use a definition of I_t which admits growing information sets, i.e. $I_t \subset I_{t+1}$. That definition of I_t enables the application of the so-called *law of iterated expectations* derived by using a standard property of conditional expectation.

Proposition 2.1 Consider an arbitrary stochastic process $\{z_t, t \in \mathcal{T}\}$ and an increasing sequence of information sets $\{I_t, t \in \mathcal{T}\}$. The expectation formed at time $t-j$, $j > 0$ of an expectation of a random variable z_{t+h}, $h > 0$, based on the information set I_t, equals the expectation of that random variable based on the information set I_{t-j}. Formally, for $\forall t$

$$E[E[z_{t+h} \mid I_t] \mid I_{t-j}] = E[z_{t+h} \mid I_{t-j}]$$

Proof of proposition 2.1 See, for instance, Shiryaev (1989) p 216 or Chow, Teicher (1988) p 208. ||

If we interpret conditional expectation as a forecast of a random variable, the law of iterated expectations says that the best forecast at time $t-j$, $j > 0$ of a forecast at time t concerning a random variable in $t+h$, $h > 0$ is the forecast formed in time $t-j$ for that random variable itself.

Based on the law of iterated expectations, the following properties can easily be derived[24].

- A rational prediction error has zero conditional mean. Formally, by applying the law of iterated expectations for $\forall t$, we get

$$E[\epsilon_{t+1} \mid I_t] = E[y_{t+1} - E[y_{t+1} \mid I_t] \mid I_t] = 0$$

  This property of rational prediction errors means that the guess about future values must be correct on average establishing the unbiasedness of rationally forecasted variables.

- A rational prediction error is uncorrelated with any element of I_t. Formally, by using a property of conditional expectation[25] for $\forall z_t \in I_t$ and $\forall t$, we get

$$\begin{aligned} E[z_t\epsilon_{t+1} \mid I_t] &= E[z_t y_{t+1} - z_t E[y_{t+1} \mid I_t] \mid I_t] \\ &= E[z_t y_{t+1} \mid I_t] - E[z_t y_{t+1} \mid I_t] = 0 \end{aligned}$$

  This property is also–called *orthogonality property* of a rational prediction error.

Both characteristics of rational prediction errors and, therefore, of rational expectations are desirable properties in econometric theory since they form the basis of optimality considerations. For, unlike the assumption of adaptive or extrapolative expectations, usually imposed ad hoc in economic modelling without further justification, it turns out that a rationally expected variable provides the best forecast among all possible forecasts based on I_t if optimality is measured in terms of the variance of the forecast error $E[(y_{t+1}-\hat{y}_{t,t+1})^2]$. Hence, the application of rational expectations may be justified by optimality reasons.

Proposition 2.2 A rationally expected variable provides an optimal forecast in the sense that by using the rationally expected variable as a forecast the variance of the forecast error becomes minimal. Formally, for $\forall t$

$$E[y_{t+1} \mid I_t] \in \{\hat{y}_{t,t+1} = f(I_t) \mid E[(y_{t+1} - \hat{y}_{t,t+1})^2] \to \min\}$$

[24]See among others Sheffrin (1983) p 7.

[25]See Malliaris, Brock (1982) p 15.

where $\hat{y}_{t,t+1}$ denotes an arbitrary forecast depending on information set I_t.

Proof of proposition 2.2 See Broze, Szafarz (1991) p 4. This proposition can easily be shown by rearranging and substituting equation (2.3) into the expression of the variance of the forecast error. Thus,

$$\begin{aligned} E[(y_{t+1} - \hat{y}_{t,t+1})^2] &= E[y_{t+1}^2 - 2y_{t+1}\hat{y}_{t,t+1} + \hat{y}_{t,t+1}^2] \\ &= E[y_{t+1}^2] - 2E[y_{t+1}\hat{y}_{t,t+1}] + E[\hat{y}_{t,t+1}^2] \end{aligned}$$

By using rational expectations hypothesis, i.e. $\hat{y}_{t,t+1} = E[y_{t+1} \mid I_t] = y_{t+1} - \epsilon_{t+1}$ and the orthogonality property, we get

$$\begin{aligned} E[(y_{t+1} - \hat{y}_{t,t+1})^2] &= E[y_{t+1}^2] - 2E[y_{t+1}(y_{t+1} - \epsilon_{t+1})] \\ &\quad + E[(y_{t+1} - \epsilon_{t+1})^2] \\ &= E[\epsilon_{t+1}^2] \end{aligned}$$

Thus, applying the rational expectations hypothesis, the variance of the forecast error equals the variance of the rational prediction error which can be regarded as an unsystematic error component. Deviations from rationally expected variables are purely random and, therefore, unpredictable. Hence, assuming rational expectations, no systematic component has any influence on $E[(y_{t+1} - \hat{y}_{t,t+1})^2]$ avoiding the disadvantage of other expectations formation processes. ||

The rationale behind this optimality property lies in the fact that individuals are assumed to include all information being relevant for forecasting within their expectation building process. No systematic error is left, errors are only due to unsystematic (erratic) deviations. This optimality primarily justifies, unlike the ad-hoc assumption of adaptive or extrapolative expectations, the use of the rational expectations hypothesis in economic modelling[26].

[26]Notice that this optimality is also the reason for rational expectations being rational. Subjective expectation is related to a sort of objective expectation in terms of conditional expectation since this expectation formation does not include any systematic error. But notice further that this concept of rationality differs from other definitions of rationality concerning subjective expectations based on an axiomatic approach for instance.

The rational expectations hypothesis usually implies two additional assumptions which play an important role in the discussion of its empirical relevance:

- In rational expectations models it is usually assumed that individuals know (or behave as if they know) the true economic model[27]. This argument has two aspects. First, economic theory must be able to provide sufficient insight into how reality works in order to build and estimate an econometric model for successfully forecasting all relevant variables. This seems to be a rather restrictive assertion seldom satisfied in the real world[28]. But we can at least suppose a learning process (of researchers, for instance) which brings economic modelling closer to reality. Second, individuals must be able to easily acquire information concerning conditional expectation (i.e. the optimal forecast) of all relevant variables. They need not necessarily know the whole underlying economic model including all equations and estimated parameters. However, they should be able to gather information concerning the conditional expectation of all relevant variables provided by researchers, for instance, who make their results publicly available.

- Furthermore, in rational expectations models, individuals are assumed to have the same information set I_t which implies two important consequences. First, individuals must agree on the underlying economic model which is implicitly contained in the information set I_t as mentioned above. This rules out any situation where different opinions regarding the economic model coexist, e.g. a Keynesian and a monetarist view of the economic world. Second, all past realizations of variables which are relevant for forming expectations are known by all individuals and can, therefore, be regarded as common

[27]Actually, the information set I_t should be enlarged by incorporating the entire economic model which is thought of as ruling the (at least, economic) world. Nevertheless, a description of the underlying model within the information set is neglected here.

[28]Shiller (1990), for instance, stresses that assuming that both the economic (or econometric) model and the popular model (i.e. the model which is applied in reality) coincide is a 'gross oversimplification' and emphasizes the importance of further research on the field of popular models.

knowledge. This rules out any situation in which insiders possibly can take advantage of any private knowledge[29].

Information homogenously distributed among individuals concerning both the model itself and the structure of the information set I_t is crucial within rational expectations models. It is usually justified by the possibility of aggregation among individuals and, therefore, the use of representative agent models which are frequently used in economic theory. However, a possible failure of rational expectations models in explaining real world phenomena may eventually be due to heterogenous information which justifies further research in this field[30].

After defining the expectational and informational framework of equation (2.4), we will now examine the solution procedure more precisely.

2.3 Solution procedure

Assuming individuals form expectations rationally, we can rewrite equation (2.1) as

$$y_t = \alpha E[y_{t+1} \mid I_t] + x_t \tag{2.4}$$

where I_t is part of a sequence of growing information sets allowing for the application of the law of iterated expectations as defined above[31]. It is

[29]Further, the information set I_t is often assumed to be stable in the sense that it includes always the same bundle of random variables at any point of time.

[30]Recently, some rational expectations models were developed explicitly taking into account heterogenous information. In particular, within a framework of efficient capital or foreign exchange markets, chartists and fundamentalists can be modelled as using different expectation formation processes. See, for instance, Frenkel, Froot (1986). Moreover, Futia (1981) provides a model where two groups interact on a market taking expectations conditional on different information sets whereas the information set of one group is a subset of the information set of the other.

[31]Equation (2.4) can be viewed as a very special case of a more general rational expectations model. Equation (2.4) takes into account only one endogenous and exogenous variable x (bivariate case), only expectations of the endogenous variable y one period ahead based on an information set of time t and only the current value of the exogenous variable x. More general models include a higher degree of lagged expectations of the

easily seen that the influence of the exogenous variable x on the endogenous variable y is twofold as shown by figure 2.1. A change in x_t affects directly

Figure 2.1: Influence of the exogenous on the endogenous variable

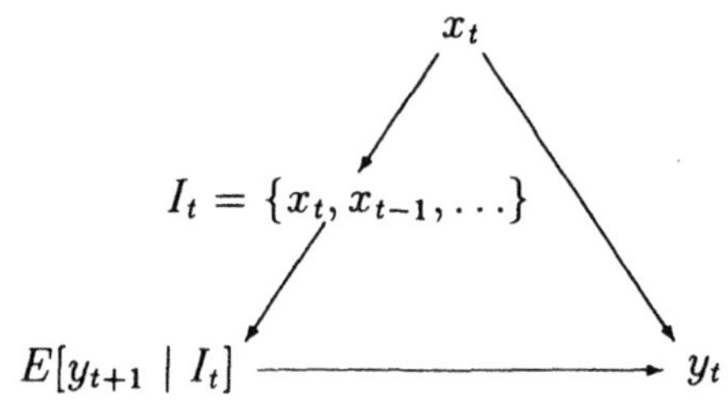

the endogenous variable y_t through being a component of equation (2.4) and indirectly through being contained in the information set I_t. This indirect influence, usually called magnification effect, may lead to a revison of the expectation formation process and may enhance the variation in the endogenous variable[32]. Additionally, the exogenous variable x is the decisive variable in the sense that it determines the stochastic nature of the model: A stochastic exogenous variable x will make the whole model stochastic; a deterministic exogenous variable x reduces the model to its perfect foresight counterpart solvable by familiar tools of deterministic difference equations.

In order to solve our rational expectations model of equation (2.4), formulated as a linear stochastic difference equation, we look at solution procedures applied to linear deterministic difference equations. Solution procedures to deterministic difference equations distinguish different solution components. It turns out that similar distinctions can be made in a stochastic context. Hence, the general solution of equation (2.4) consists of two components: a particular solution and a complementary solution.

---

endogenous variable y based on differently lagged information sets and lagged values of the exogenous variable x. In addition, the number of both endogenous and exogenous variables may be increased (multivariate case). See Broze, Szafarz (1991) for a detailed analysis of solution procedures of more general rational expectations models.

[32]See Gaab (1990) and Heri (1986), for instance.

2.3.1 Particular solution

In the case of the particular solution different solution procedures can be separated depending on whether a successive substitution procedure (forward or backward solution) without further characterization of the exogenous stochastic process $\{x_t, t \in \mathcal{T}\}$ is considered, or whether we assume the exogenous variable x to follow a special stochastic process. Concerning the former more general procedure, a distinction between two cases depending on the value taken by the parameter α may be useful: for $|\alpha| > 1$ the backward solution can be derived while the forward solution is obtained for $|\alpha| < 1$. Concerning the latter procedure, two stochastic processes for the exogenous variable x will be considered and the corresponding particular solutions derived, i.e. a stationary ARMA–process and a non–stationary random walk.

2.3.1.1 Forward and backward solution

Solving equation (2.4) without assuming a special stochastic process for the exogenous variable x, we are able to derive forward and backward solutions. The idea is to eliminate the conditional expectation in equation (2.4) by successive (forward or backward looking) substitution[33].

Starting with the *backward solution*, we try to express the endogenous stochastic process $\{y_t, t \in \mathcal{T}\}$ through past and current values of the exogenous variable x. This is only possible if we introduce an additional assumption. Blanchard (1979a) proposed the perfect foresight assumption claiming that the expectation of a random variable equals its true value, i.e. $E[y_{t+1} \mid I_t] = y_{t+1}$. Notice that this is a rather simple solution to the problem of characterizing expectations of future variables by assuming that the true value of the endogenous variable y is perfectly known at time t. The assertion of perfect foresight may be justified by the so–called 'certainty equivalence' of rational expectations, which means that expected values differ from actual values only randomly with zero deviation on average captured above by the mean zero property of rational prediction

[33]See among others Blanchard (1979a) and Blanchard, Kahn (1980). For the continuous counterpart see Ikeda, Shibata (1992).

errors. Accepting this, assumption equation (2.4) becomes

$$y_t = \alpha y_{t+1} + x_t$$

or after lagging once

$$y_t = \frac{1}{\alpha} y_{t-1} - \frac{1}{\alpha} x_{t-1}$$

we obtain a difference equation which can easily be solved by recursive substitution of prior realizations of the endogenous variable y up to time $t - T$.

$$y_t = \frac{1}{\alpha^T} y_{t-T} - \left[\frac{1}{\alpha} x_{t-1} + \frac{1}{\alpha^2} x_{t-2} + \ldots + \frac{1}{\alpha^T} x_{t-T} \right]$$

By extension ($T \to \infty$), the first term on the right side approaches zero if

(i) $y_{-\infty} < \infty$

(ii) $|\alpha| > 1$

Condition (i) depends on the stochastic nature of x_t since the endogenous variable is exclusively driven by the exogenous variable. Condition (ii) is the familiar stability condition of linear difference equations. The resulting expression (2.5) is called the backward looking particular solution of equation (2.4).

$$y_t = - \sum_{i=1}^{\infty} \frac{1}{\alpha^i} x_{t-i} \tag{2.5}$$

The endogenous stochastic process $\{y_t, t \in \mathcal{T}\}$ is fully determined by an infinite sum consisting of past values of the exogenous variable x, each element x_{t-i} of the sum weighted by the factor $\frac{1}{\alpha^i}$. Notice that the realization of the exogenous stochastic process at time t has no influence on y_t. The backward solution (2.5) exists if

(iii) the exogenous stochastic process $\{x_t, t \in \mathcal{T}\}$ does not grow or shrink (depending on the sign of x_t) too fast in comparison to the increase of the weights[34]

(iv) $|\alpha| > 1$

[34]Some authors postulate $\{x_t, t \in \mathcal{T}\}$ to be stationary, however, this is unnecessarily restrictive. This weaker condition enables us to consider non–stationary exogenous processes below, i.e. random walks.

Both conditions guarantee the convergence of the sum in equation (2.5).

In summary, the exogenous stochastic process $\{x_t, t \in \mathcal{T}\}$ considered from the beginning up to time $t-1$ defines under the assumption of perfect foresight and condition (i)–(iii) the particular solution of equation (2.4). The backward looking sequence of the exogenous variable x names this particular solution.

Consider now the *forward solution*. In order to remove the conditional expectation in equation (2.4), the expectational term is successively substituted by alternative expressions containing expectations of the endogenous variable y in later periods. Leading once and taking expectations of equation (2.4) based on the current information set I_t we get

$$E[y_{t+1} \mid I_t] = \alpha E[E[y_{t+2} \mid I_{t+1}] \mid I_t] + E[x_{t+1} \mid I_t]$$

Applying the law of iterated expectations to the first term on the right side, we obtain

$$E[y_{t+1} \mid I_t] = \alpha E[y_{t+2} \mid I_t] + E[x_{t+1} \mid I_t]$$

By successive substitution of expected values of the endogenous variable y up to period T, equation (2.4) can be rewritten as

$$\begin{aligned} y_t &= \alpha E[y_{t+1} \mid I_t] + x_t \\ &= \alpha^2 E[y_{t+2} \mid I_t] + \alpha E[x_{t+1} \mid I_t] + x_t \\ &\dots \\ &= \alpha^{T+1} E[y_{t+T+1} \mid I_t] + \alpha^T E[x_{t+T} \mid I_t] + \ldots + \alpha E[x_{t+1} \mid I_t] + x_t \end{aligned}$$

or equivalently

$$y_t = \alpha^{T+1} E[y_{t+T+1} \mid I_t] + \sum_{i=0}^{T} \alpha^i E[x_{t+i} \mid I_t]$$

Again, by extension ($T \to \infty$), the first term on the right side converges to zero, i.e.

$$\lim_{T \to \infty} \alpha^{T+1} E[y_{t+T+1} \mid I_t] = 0 \tag{2.6}$$

if

(i) $y_\infty < \infty$

(ii) $|\alpha| < 1$

Condition (2.6) is usually called transversality condition. The resulting expression (2.7) represents the forward looking particular solution of equation (2.4).

$$y_t = \sum_{i=0}^{\infty} \alpha^i E[x_{t+i} \mid I_t] \tag{2.7}$$

The endogenous stochastic process $\{y_t, t \in \mathcal{T}\}$ is fully determined by an infinite sum consisting of expected future values of the exogenous variable x, each element x_{t+i} of the sum discounted by α^i. The forward solution (2.7) exists if

(iii) the exogenous stochastic process $\{x_t, t \in \mathcal{T}\}$ does not grow or shrink (depending on the sign of x_t) too fast in comparison to the decrease of the weights[35]

(iv) $|\alpha| < 1$

Both conditions guarantee the convergence of the sum in equation (2.7)[36].

In summary, the forward solution consists of conditional expectations of the exogenous variable x from now on up to infinity. It exists under the assumption of growing information sets (the law of iterated expectations is applicable) and condition (i)–(iii). This forward looking sequence of expected values of the exogenous variable x names this particular solution.

To conclude, we have derived rather general particular solutions by backward and forward substitution. It turned out that for $|\alpha| > 1$ the backward looking procedure, for $|\alpha| < 1$ the forward looking procedure is obtained.

2.3.1.2 Explicit particular solutions

Besides the solution procedure of forward or backward looking substitution, explicit solutions can be derived by assuming special stochastic processes for the exogenous variable x since the infinite sum of forward and

[35]See footnote 34.

[36]Below we derive explicit restrictions on the parameter involved in order to assure the existence of the forward solution.

backward solution is now solvable. Here, two stochastic processes will be considered: first a parametric stationary ARMA–process and second a non–parametric non–stationary random walk.

1. ARMA–solutions

One of the probably most applied assumptions concerning the exogenous stochastic process $\{x_t, t \in \mathcal{T}\}$ is that it follows an autoregressive moving–average (ARMA) process.

Definition 2.4 Consider a stochastic process $\{x_t, t \in \mathcal{T}\}$. If there exist real numbers $\phi_1, \ldots, \phi_p, \theta_1, \ldots, \theta_q$ and a white noise ϵ_t such that

$$x_t + \phi_1 x_{t-1} + \ldots + \phi_p x_{t-p} = \epsilon_t + \theta_1 \epsilon_{t-1} + \ldots + \theta_q \epsilon_{t-q}$$

with $\phi_p \neq 0$, $\theta_p \neq 0$ and the polynominals

$$\begin{aligned} \Phi(u) &= 1 + \phi_1 u + \ldots + \phi_p u^p \\ \Theta(u) &= 1 + \theta_1 u + \ldots + \theta_q u^q \end{aligned}$$

have no common roots, then $\{x_t, t \in \mathcal{T}\}$ can be represented by an *autoregressive moving–average process* of order (p, q) (ARMA(p, q)).

For notational convenience let L denote a lag operator with $Lx_t = x_{t-1}$. An ARMA–process can then equivalently be expressed as

$$\Phi(L)x_t = \Theta(L)\epsilon_t \tag{2.8}$$

with $\Phi(L) = 1 + \phi_1 L + \ldots + \phi_p L^p$ and $\Theta(L) = 1 + \theta_1 L + \ldots + \theta_q L^q$. The particular solution can be derived explicitly when fundamentals follow an ARMA–process, as has been proven by Gourieroux, Laffont and Monfort (1982)[37].

Proposition 2.3 If the exogenous stochastic process $\{x_t, t \in \mathcal{T}\}$ can be described as an ARMA–process, then equation (2.4) has the ARMA–solution

$$\Theta(L)y_t = \Gamma(L)x_t$$

[37] Other references calculating ARMA–solutions are Evans, Honkapohja (1986), Evans (1987) and d'Autumne (1990).

where $\Theta(L)$ is the polynomial defined in (2.8) and the polynomial $\Gamma(L)$ can be expressed as

$$\Gamma(L) = \frac{1}{L-\alpha}\left(L\Theta(L) - \alpha\frac{\Theta(\alpha)}{\Phi(\alpha)}\Phi(L)\right)$$

Proof of proposition 2.3 See Gourieroux, Laffont and Monfort (1982) proving this proposition rather shortly. Here, a more detailed version is provided. The solution procedure relies on the method of undetermined coefficients which consists of a guess of the form of the solution. Suppose the exogenous stochastic process $\{x_t, t \in \mathcal{T}\}$ follows an ARMA–process, i.e.

$$\Phi(L)x_t = \Theta(L)\epsilon_t \tag{2.9}$$

where $\Phi(L)$ and $\Theta(L)$ are defined above. Our guess concerning the ARMA–solution may have the following form

$$\Theta(L)y_t = \Gamma(L)x_t \tag{2.10}$$

where $\Gamma(L)$ denotes the polynom $\Gamma(L) = \gamma_0 + \gamma_1 L + \gamma_2 L^2 \ldots$. The proof follows four steps.

First, expressing $E[x_{t+1} \mid I_t]$ in terms of x_t. From assumption (2.9) and under the condition of rational expectations we obtain

$$\begin{aligned}
x_{t+1} &= \frac{\Theta(L)}{\Phi(L)}\epsilon_{t+1} \\
&= \frac{\Theta(L)}{\Phi(L)}(x_{t+1} - E[x_{t+1} \mid I_t]) \\
\left(1 - \frac{\Theta(L)}{\Phi(L)}\right)\frac{1}{L}x_t &= -\frac{\Theta(L)}{\Phi(L)}E[x_{t+1} \mid I_t] \\
E[x_{t+1} \mid I_t] &= \left(1 - \frac{\Phi(L)}{\Theta(L)}\right)\frac{1}{L}x_t \qquad (2.11)
\end{aligned}$$

Second, expressing $E[y_{t+1} \mid I_t]$ in terms of x_t. From assumption (2.10) we obtain

$$\begin{aligned}
\Theta(L)y_{t+1} &= \Gamma(L)x_{t+1} \\
&= \gamma_0 x_{t+1} + \gamma_1 x_t + \gamma_2 x_{t-1} + \ldots
\end{aligned}$$

$$\begin{aligned}
\Theta(L)E[y_{t+1} \mid I_t] &= \gamma_0 E[x_{t+1} \mid I_t] + \gamma_1 x_t + \gamma_2 x_{t-1} + \dots \\
&\quad + \gamma_0 x_{t+1} - \gamma_0 x_{t+1} \\
&= \gamma_0 E[x_{t+1} \mid I_t] + (\Gamma(L) - \gamma_0)\frac{1}{L}x_t \\
E[y_{t+1} \mid I_t] &= \frac{\gamma_0}{\Theta(L)} E[x_{t+1} \mid I_t] + \left(\frac{\Gamma(L)}{\Theta(L)} - \frac{\gamma_0}{\Theta(L)}\right)\frac{1}{L}x_t \\
(2.12) \quad E[y_{t+1} \mid I_t] &= \gamma_0 E[x_{t+1} \mid I_t] + \left(\frac{\Gamma(L)}{\Theta(L)} - \gamma_0\right)\frac{1}{L}x_t
\end{aligned}$$

The last rearrangement is due to the fact that all realizations of the first and the third term prior to time $t+1$ on the right side are equal and, therefore, cancel out.

Third, substitution of equation (2.11) and equation (2.12) into equation (2.4). After calculating conditional expectation terms in terms of x_t, we can solve equation (2.4) by the method of undetermined coefficients. Taking our guess (2.10) into consideration and substituting equation (2.11) and equation (2.12) into equation (2.4), we obtain

$$\frac{\Gamma(L)}{\Theta(L)}x_t = \alpha\left[\gamma_0\left\{\left(1 - \frac{\Phi(L)}{\Theta(L)}\right)\frac{1}{L}x_t\right\} + \left(\frac{\Gamma(L)}{\Theta(L)} - \gamma_0\right)\frac{1}{L}x_t\right] + x_t$$

Simple rearrangement yields

$$(2.13) \qquad L\Gamma(L) = -\alpha\gamma_0\Phi(L) + \alpha\Gamma(L) + L\Theta(L)$$

Fourth, determination of the unknown coefficients. By taking $L = \alpha$ in equation (2.13), the term on the left side and the second term on the right side cancel out. We obtain for γ_0 ($\Phi(\alpha) \neq 0$)

$$(2.14) \qquad \gamma_0 = \frac{\Theta(\alpha)}{\Phi(\alpha)}$$

Substituting γ_0 in equation (2.13) and solving for $\Gamma(L)$ results the desired ARMA–solution

$$\Gamma(L) = \frac{1}{L-\alpha}\left(L\Theta(L) - \alpha\frac{\Theta(\alpha)}{\Phi(\alpha)}\Phi(L)\right)$$

Obviously, for $L = 0$ which corresponds to γ_0 we get the necessary condition of equation (2.14). ||

Notice that the ARMA–solution is derived regardless of the value taken by the parameter α. The only restriction imposed is $\Phi(\alpha) \neq 0$[38].

Example: An example may be useful here. Suppose the exogenous process can be described by an AR(1)–process, i.e. $x_t = \phi x_{t-1} + \epsilon_t$ with $|\phi| < 1$. According to proposition 2.3, we obtain for $\Theta(L) = 1$ and $\Phi(L) = 1 - \phi L$ the following particular solution

$$y_t = \frac{1}{L-\alpha}\left(L - \alpha\frac{1}{1-\phi\alpha}(1-\phi L)\right)x_t = \frac{1}{1-\phi\alpha}x_t$$

Hence, the current realization of the endogenous stochastic process y_t is proportional to the current value of the exogenous stochastic process x_t if $\{x_t, t \in \mathcal{T}\}$ can be described as an ARMA–process.

A natural generalization of the above proposition may consist in taking into account a drift parameter μ[39] which is included in many econometric models[40], i.e. the exogenous stochastic process $\{x_t, t \in \mathcal{T}\}$ can be described by

$$\Phi(L)x_t = \mu + \Theta(L)\epsilon_t$$

The following proposition computes the ARMA–solution of equation (2.4) when the exogenous stochastic process $\{x_t, t \in \mathcal{T}\}$ contains a drift parameter.

Proposition 2.4 If the exogenous stochastic process $\{x_t, t \in \mathcal{T}\}$ can be described as an ARMA–process with drift, equation (2.4) has the ARMA–solution

$$\Theta(L)y_t = \Gamma(L)x_t + \frac{\alpha}{1-\alpha}\frac{\Theta(\alpha)}{\Phi(\alpha)}\mu$$

---

[38] Using the relation between the exogenous variable x and the error term expressed in equation (2.8), the ARMA–solution can equivalently be expressed in terms of the white noise ϵ_t by rearranging equation (2.9) and (2.10), yielding the moving–average representation of the ARMA–solution $y_t = \frac{1}{L-\alpha}\left(L\frac{\Theta(L)}{\Phi(L)} - \alpha\frac{\Theta(\alpha)}{\Phi(\alpha)}\right)\epsilon_t$.

[39] Notice that considering a drift component μ represents the only possibility of generalization. An additional trend component would violate the stationarity property of ARMA–processes.

[40] Alternatively, the exogenous stochastic process may be transformed by reformulation of $\{x_t, t \in \mathcal{T}\}$ in deviations from its mean. Then, subsequent application of proposition 2.3 is possible.

where $\Theta(L)$ and $\Phi(L)$ are the polynomials defined in (2.8), μ represents the drift parameter and the polynomial $\Gamma(L)$ can be expressed as

$$\Gamma(L) = \frac{1}{L-\alpha}\left(L\Theta(L) - \alpha\frac{\Theta(\alpha)}{\Phi(\alpha)}\Phi(L)\right)$$

Proof of proposition 2.4 We extend the above procedure applied by Gourieroux, Laffont and Monfort (1982) by including a drift component. Thus, suppose the exogenous stochastic process $\{x_t, t \in \mathcal{T}\}$ follows an ARMA–process with drift, i.e.

$$\Phi(L)x_t = \mu + \Theta(L)\epsilon_t \tag{2.15}$$

where $\Theta(L)$ and $\Phi(L)$ are defined above. Our guess concerning the ARMA–solution may have the following form

$$\Theta(L)y_t = \Gamma(L)x_t + \delta_0\mu \tag{2.16}$$

where $\Gamma(L)$ denotes the following polynom $\Gamma(L) = \gamma_0 + \gamma_1 L + \gamma_2 L^2 \ldots$ and δ_0 an arbitrary coefficient to be determined. The proof follows four steps.

First, expressing $E[x_{t+1} \mid I_t]$ in terms of x_t. From assumption (2.15) we obtain

$$\begin{aligned}
x_{t+1} &= \frac{\mu}{\Phi(L)} + \frac{\Theta(L)}{\Phi(L)}\epsilon_{t+1} \\
&= \frac{\mu}{\Phi(L)} + \frac{\Theta(L)}{\Phi(L)}(x_{t+1} - E[x_{t+1} \mid I_t]) \\
\left(1 - \frac{\Theta(L)}{\Phi(L)}\right)\frac{1}{L}x_t &= \frac{\mu}{\Phi(L)} - \frac{\Theta(L)}{\Phi(L)}E[x_{t+1} \mid I_t] \\
E[x_{t+1} \mid I_t] &= \frac{\mu}{\Theta(L)} + \left(1 - \frac{\Phi(L)}{\Theta(L)}\right)\frac{1}{L}x_t \qquad (2.17)
\end{aligned}$$

Second, expressing $E[y_{t+1} \mid I_t]$ in terms of x_t. From assumption (2.16) we obtain

$$\begin{aligned}
\Theta(L)y_{t+1} &= \Gamma(L)x_{t+1} + \delta_0\mu \\
&= \gamma_0 x_{t+1} + \gamma_1 x_t + \gamma_2 x_{t-1} + \ldots + \delta_0\mu \\
\Theta(L)E[y_{t+1} \mid I_t] &= \gamma_0 E[x_{t+1} \mid I_t] + \gamma_1 x_t + \gamma_2 x_{t-1} + \ldots
\end{aligned}$$

$$
\begin{aligned}
& +\gamma_0 x_{t+1} - \gamma_0 x_{t+1} + \delta_0\mu \\
&= \gamma_0 E[x_{t+1} \mid I_t] + (\Gamma(L) - \gamma_0)\frac{1}{L}x_t + \delta_0\mu
\end{aligned}
$$

$$
(2.18)\quad E[y_{t+1} \mid I_t] = \gamma_0 E[x_{t+1} \mid I_t] + \left(\frac{\Gamma(L)}{\Theta(L)} - \gamma_0\right)\frac{1}{L}x_t + \frac{\delta_0\mu}{\Theta(L)}
$$

Third, substitution of equation (2.17) and equation (2.18) into equation (2.4). After calculating conditional expectation in terms of x_t, we can solve equation (2.4) by the method of undetermined coefficients. Substitution of equation (2.17) and equation (2.18) into equation (2.4) yields

$$
\begin{aligned}
(2.19)\quad \frac{\Gamma(L)}{\Theta(L)}x_t + \frac{\delta_0\mu}{\Theta(L)} &= \alpha\left[\gamma_0\left\{\left(1 - \frac{\Phi(L)}{\Theta(L)}\right)\frac{1}{L}x_t + \frac{\mu}{\Theta(L)}\right\}\right. \\
&\quad + \left.\left(\frac{\Gamma(L)}{\Theta(L)} - \gamma_0\right)\frac{1}{L}x_t + \frac{\delta_0\mu}{\Theta(L)}\right] + x_t
\end{aligned}
$$

Fourth, determination of the unknown coefficient δ_0. Since the drift parameter μ appears in only three terms of equation (2.19), we can easily obtain an expression for δ_0 by the method of undetermined coefficients taking into account only those components of equation (2.19) which contain δ_0.

$$
\begin{aligned}
\frac{\delta_0}{\Theta(L)}\mu &= \frac{\alpha\gamma_0}{\Theta(L)}\mu + \frac{\alpha\delta_0}{\Theta(L)}\mu \\
\delta_0 &= \frac{\alpha}{1-\alpha}\gamma_0
\end{aligned}
$$

It turns out that δ_0 is proportional to the coefficient γ_0. Since the determination of the parameter γ_0 is independent of the determination of the parameter δ_0, the above analysis applies. In particular, we obtain the same expressions for γ_0 and $\Gamma(L)$. Substituting $\Gamma(L)$, γ_0 and δ_0 in equation (2.16) we get

$$
\Theta(L)y_t = \Gamma(L)x_t + \frac{\alpha}{1-\alpha}\frac{\Theta(\alpha)}{\Phi(\alpha)}\mu
$$

which is the desired particular solution if x_t follows an ARMA–process with drift μ.

||

Example: Again, an example may be useful here. Suppose the exogenous process can be described by an AR(1)-process with drift, i.e. $x_t = \mu + \phi x_{t-1} + \epsilon_t$ with $|\phi| < 1$. By applying proposition 2.4, the fundamental solution is

$$y_t = \frac{1}{1-\phi\alpha}x_t + \frac{\alpha}{(1-\alpha)(1-\phi\alpha)}\mu$$

Hence, the endogenous stochastic process $\{y_t, t \in \mathcal{T}\}$ is a linear combination of the exogenous stochastic process $\{x_t, t \in \mathcal{T}\}$.

Recall that the ARMA–solution derived by proposition 2.3 and proposition 2.4 exists regardless of the value taken by the parameter α. However, we can derive a broader ARMA–solution set under the restricting assumption of $|\alpha| > 1$.

Proposition 2.5 If the exogenous stochastic process $\{x_t, t \in \mathcal{T}\}$ can be described as an ARMA–process and $|\alpha| > 1$, then equation (2.4) has the ARMA–solution

$$\Theta(L)y_t = \Delta(L)x_t$$

where $\Theta(L)$ is the polynomial defined in (2.8) and the polynomial $\Delta(L)$ can be expressed as

$$\Delta(L) = \frac{1}{L-\alpha}(L\Theta(L) - \alpha c_0 \Phi(L))$$

where c_0 is a real parameter.

Proof of proposition 2.5 See Broze, Szafarz (1991) p 34. ||

Notice that for $c_0 = \frac{\Theta(\alpha)}{\Phi(\alpha)}$ the ARMA–solution of proposition 2.3 coincides with the ARMA–solution of proposition 2.5 characterizing the former as a subset of the latter in the case of $|\alpha| > 1$. Hence, unlike the case $|\alpha| < 1$, where the ARMA–solution can uniquely be defined, the existence of an arbitrary parameter c_0 establishes a multiplicity of the ARMA–solutions in the case of $|\alpha| > 1$. This seems to be curious since we observed no such multiplicity when deriving particular solutions more generally by repeated substitution. Therefore, an investigation of how both solution sets are related might be interesting.

Relationship between ARMA– and forward and backward solutions

So far, we derived the particular solution of equation (2.4) through successive substitution allowing for an arbitrary exogenous stochastic process $\{x_t, t \in \mathcal{T}\}$ and the particular solutions of ARMA–representable exogenous stochastic processes $\{x_t, t \in \mathcal{T}\}$. It may be of further interest to see how these solution sets are connected.

Thus, suppose the exogenous stochastic process $\{x_t, t \in \mathcal{T}\}$ is representable by an ARMA–process. Following the distinction made above when deriving backward and forward solutions, we can discriminate between the two cases $|\alpha| < 1$ and $|\alpha| > 1$.

CASE 1: $|\alpha| < 1$

For $|\alpha| < 1$ only proposition 2.3 is applicable. We can show that the solution of proposition 2.3 equals the forward solution.

Proposition 2.6 If the exogenous stochastic process $\{x_t, t \in \mathcal{T}\}$ can be described as an ARMA–process and $|\alpha| < 1$ holds, then the unique ARMA–solution of proposition 2.3 corresponds to the forward solution.

Proof of proposition 2.6 See Broze, Szafarz (1991) p 34. Here, we provide a more extended version of the proof. According to equation (2.7), the forward solution can be expressed as

$$y_t = \sum_{i=0}^{\infty} \alpha^i E[x_{t+i} \mid I_t]$$

Admitting an ARMA–representation for the exogenous stochastic process $\{x_t, t \in \mathcal{T}\}$ and using the lag operator L we can rewrite the forward solution as

$$y_t = \sum_{i=0}^{\infty} \alpha^i E\left[\frac{\Theta(L)}{\Phi(L)} L^{-i} \epsilon_t \mid I_t\right] \tag{2.20}$$

where negative powers of the lag operator L denotes future values of ϵ_t. In order to consider future values of ϵ_t, define the following polynomial

$$\frac{\Theta(L)}{\Phi(L)} = A(L) = a_0 + a_1 L + a_2 L^2 + \dots \quad \text{with } a_0 = 1$$

Substituting this expression in equation (2.20) we obtain by taking expectations

$$y_t = \left(\sum_{i=0}^{\infty} \alpha^i A(L) L^{-i} - \sum_{i=1}^{\infty} \sum_{j=0}^{i-1} \alpha^i L^{-i} a_j L^j \right) \epsilon_t \tag{2.21}$$

Notice that $E[\epsilon_{t+i} \mid I_t] = 0$ for $\forall i > 1$. The first infinite sum in equation (2.21) can easily be solved.

$$\sum_{i=0}^{\infty} \alpha^i A(L) L^{-i} = \frac{A(L)}{1 - \frac{\alpha}{L}} = \frac{LA(L)}{L - \alpha} \tag{2.22}$$

The second double sum in equation (2.21) is a bit harder to solve. Consider the following decomposition of the double sum

$$\begin{aligned}
\sum_{i=1}^{\infty} \sum_{j=0}^{i-1} \alpha^i L^{-i} a_j L^j &= \alpha L^{-1} a_0 \\
&\quad + \alpha^2 L^{-2}(a_0 + a_1 L) \\
&\quad + \alpha^3 L^{-3}(a_0 + a_1 L + a_2 L^2) \\
&\quad + \dots \\
&= a_0(\alpha L^{-1} + \alpha^2 L^{-2} + \alpha^3 L^{-3} + \dots) \\
&\quad + a_1 L(\alpha^2 L^{-2} + \alpha^3 L^{-3} + \dots) \\
&\quad + a_2 L^2(\alpha^3 L^{-3} + \dots) \\
&\quad + \dots \\
&= a_0(\alpha L^{-1})(1 + \alpha L^{-1} + \dots) \\
&\quad + a_1 \alpha(\alpha L^{-1})(1 + \alpha L^{-1} + \dots) \\
&\quad + a_2 \alpha^2(\alpha L^{-1})(1 + \alpha L^{-1} + \dots) \\
&\quad + \dots
\end{aligned}$$

The last expression can equivalently be written in terms of an infinite sum as

$$= \frac{\alpha L^{-1}}{1 - \alpha L^{-1}} \sum_{j=0}^{\infty} \alpha^j a_j \tag{2.23}$$

By using the definitorial polynom $A(L)$, we can express the infinite sum in equation (2.23) as

$$\sum_{j=0}^{\infty} \alpha^j a_j = \frac{\Theta(\alpha)}{\Phi(\alpha)}$$

Hence, the double sum in equation (2.21) equals the following expression.

$$\sum_{i=1}^{\infty}\sum_{j=0}^{i-1} \alpha^i L^{-i} a_j L^j = \frac{\alpha}{L-\alpha}\frac{\Theta(\alpha)}{\Phi(\alpha)} \tag{2.24}$$

Substitution of the simplified expressions of equation (2.22) and (2.24) for the infinite sums into equation (2.21) yields the desired result. ||

CASE 2: $|\alpha| > 1$

For $|\alpha| > 1$ we have derived an infinite set of solutions because of the arbitrary real parameter c_0 in proposition 2.5. Since proposition 2.3 holds for all values of the parameter α, the forward solution is compatible with the ARMA–solution set of proposition 2.5. As mentioned earlier, the corresponding value of the arbitrary parameter is $c_0 = \frac{\Theta(\alpha)}{\Phi(\alpha)}$. Therefore, the forward solution derived from proposition 2.3 is also a solution for the case $|\alpha| > 1$. Notice that this result contradicts the analysis within the context of backward and forward looking solution procedures. There, the forward solution was derived only for $|\alpha| < 1$. Admitting an ARMA–representation for the exogenous stochastic process $\{x_t, t \in \mathcal{T}\}$ the forward solution forms part of the solution set regardless of the value taken by the parameter α. Hence, in the case of an ARMA–representable exogenous stochastic process $\{x_t, t \in \mathcal{T}\}$ the solution procedure of forward looking substitution overlooks a part of the solution set by restricting the parameter α unnecessarily strongly. Within the class of ARMA–particular solutions proposition 2.3 provides a broader solution set.

On the other side, it turned out that for $|\alpha| > 1$, the backward solution is a particular solution. It can easily be shown that for $c_0 = 0$, both the backward solution and the ARMA–solution of proposition 2.5 coincide.

Proposition 2.7 Consider the ARMA–solution set suggested by proposition 2.5. For $c_0 = 0$ the ARMA–solution equals the backward solution.

Proof of proposition 2.7 See Broze, Szafarz (1991) p 36. According to equation (2.5), the backward solution can be expressed as

$$y_t = -\sum_{i=1}^{\infty} \frac{1}{\alpha^i} x_{t-i}$$

Admitting an ARMA–representation for the exogenous stochastic process $\{x_t, t \in \mathcal{T}\}$ and using the lag operator L we can rewrite the backward solution as

$$\begin{aligned} y_t &= -\sum_{i=1}^{\infty} \frac{1}{\alpha^i} \frac{\Theta(L)}{\Phi(L)} L^i \epsilon_t \\ &= -\frac{\Theta(L)}{\Phi(L)} \sum_{i=1}^{\infty} \frac{1}{\alpha^i} L^i \epsilon_t \\ &= -\frac{\Theta(L)}{\Phi(L)} \frac{L}{\alpha - L} \epsilon_t \\ &= \frac{L}{L - \alpha} x_t \end{aligned}$$

Comparing the last expression with the ARMA–solution set described in proposition 2.5 it turns out that both coincide for $c_0 = 0$.

||

Since both the forward and the backward solution are contained in the infinite ARMA–solution set for $|\alpha| > 1$, we can extend proposition 2.5 in the following way.

Proposition 2.8 If the exogenous stochastic process $\{x_t, t \in \mathcal{T}\}$ can be described by an ARMA–process and $|\alpha| > 1$, then equation (2.4) has the solution

$$y_t = \lambda y_t^f + (1 - \lambda) y_t^b$$

where

- $0 \leq \lambda \leq 1$ is a real parameter

- $y_t^f = \frac{1}{L-\alpha}\left(L - \alpha\frac{\Theta(\alpha)}{\Phi(\alpha)}\frac{\Phi(L)}{\Theta(L)}\right)x_t$ is the forward solution.
- $y_t^b = \frac{L}{L-\alpha}x_t$ is the backward solution.

Proof of proposition 2.8 See Broze, Szafarz (1991) p 36. We use an alternative parametrization in comparison to proposition 2.5 by relating both c_0 and $0 < \lambda < 1$ through

$$c_0 = \lambda\frac{\Theta(\alpha)}{\Phi(\alpha)} \tag{2.25}$$

By substituting equation (2.25) into the ARMA–solution set of proposition 2.5, we obtain

$$\Theta(L)y_t = \frac{1}{L-\alpha}\left(L\Theta(L) - \alpha\lambda\frac{\Theta(\alpha)}{\Phi(\alpha)}\Phi(L)\right)x_t$$

Obviously, we can derive the result of proposition 2.8 as follows

$$\begin{aligned}
y_t &= \frac{1}{L-\alpha}\left(L - \alpha\lambda\frac{\Theta(\alpha)}{\Phi(\alpha)}\frac{\Phi(L)}{\Theta(L)}\right)x_t \\
&= \frac{L}{L-\alpha}x_t - \frac{\alpha}{L-\alpha}\lambda\frac{\Theta(\alpha)}{\Phi(\alpha)}\frac{\Phi(L)}{\Theta(L)}x_t \\
&= \frac{L}{L-\alpha}x_t - \lambda\frac{L}{L-\alpha}x_t + \lambda\frac{L}{L-\alpha}x_t - \frac{\alpha}{L-\alpha}\lambda\frac{\Theta(\alpha)}{\Phi(\alpha)}\frac{\Phi(L)}{\Theta(L)}x_t \\
&= \lambda y^f + (1-\lambda)y^b
\end{aligned}$$

where y^f and y^b are defined above. ||

After clarifying the relationship between ARMA– and forward and backward solutions, now the set of explicit particular solutions will be discussed where the exogenous process follows a random walk.

2. Random walk solutions

Since many economic time series involve non–stationarity, it appears to be justified to consider an alternative set of solutions besides the stationary ARMA–solution set. Here, the particular solution for exogenous stochastic processes that can be described by a non–stationary random walk will be examined. Two different random walk specifications are considered. First, a random walk in levels including drift and trend component and, second, a random walk in logs including a drift component which is usually called a geometric random walk.

In the following only the forward looking solution is established for two reasons. First, in an economic framework often only the forward solution is relevant. Second, the calculation of backward solutions is very tedious and the computational effort stands in no relation to the result which is very similar to the forward solution. Hence, we feel justified to skip the derivation of the backward solution.

Consider first a *random walk in levels with drift and trend component* which can be expressed by

$$x_t = \mu + \beta t + x_{t-1} + \epsilon_t \tag{2.26}$$

where μ denotes the drift, βt refers to the trend component and ϵ_t represents an NIID–error term[41]. Then, we can state the following proposition using the procedure applied for deriving the forward solution.

Proposition 2.9 If the exogenous stochastic process $\{x_t, t \in \mathcal{T}\}$ can be described as a random walk with drift and trend component, then equation (2.4) has the *forward solution*

$$y_t = \frac{1}{1-\alpha}x_t + \frac{\alpha}{(1-\alpha)^2}\left(\mu + \frac{1}{1-\alpha}\beta\right) + \frac{\alpha}{(1-\alpha)^2}\beta t$$

where μ denotes the drift parameter and βt the trend component.

Proof of proposition 2.9 Recall the forward solution of equation (2.4), i.e. $y_t = \sum_{i=0}^{\infty} \alpha^i E[x_{t+i} \mid I_t]$. By successive substitution of equation (2.26) into the forward solution and by taking expectations,

[41] Recall that a random walk is the discrete counterpart of a continous Brownian motion. A Brownian motion (with drift and trend component) can be expressed as $x_t = mt + bt^2 + w_t$ where w_t denotes a simple Brownian motion with independently and normally distributed increments. By differencing once we obtain $x_t = m - b + 2bt - x_{t-1} + \epsilon_t$ where $\epsilon_t = w_t - w_{t-1} \sim N(0, \sigma^2)\ \forall t$. This expression equals equation (2.26) for $\mu = m - b$ and $\beta = 2b$. For a more extended description of Brownian motions see Resnik (1992) or Karlin, Taylor (1976).

we obtain

$$
\begin{aligned}
y_t = x_t \quad & + \quad \alpha(\mu + \beta(t+1) + x_t) \\
& + \quad \alpha^2(2\mu + \beta(t+1) + \beta(t+2) + x_t) + \dots \\
= x_t \quad & + \quad \alpha x_t + \alpha^2 x_t + \dots \\
& + \quad \alpha\mu + \alpha^2 2\mu + \dots \\
& + \quad \alpha\beta(t+1) + \alpha^2(\beta(t+1) + \beta(t+2)) + \dots
\end{aligned} \tag{2.27}
$$

It can easily be seen that the first infinite sequence converges to $\frac{1}{1-\alpha}x_t$. The second infinite series can equivalently be expressed as

$$\mu\sum_{i=1}^{\infty}\alpha^i i = \mu\sum_{i=0}^{\infty}\alpha^i i = \mu\left(\sum_{i=0}^{\infty}\alpha^i(i+1) - \sum_{i=0}^{\infty}\alpha^i\right)$$

The first sum can be simplified to

$$
\begin{aligned}
\sum_{i=0}^{\infty}\alpha^i(i+1) &= 1\alpha^0 + 2\alpha^1 + 3\alpha^2 + \dots \\
&= 1 + \alpha + \alpha + \alpha^2 + \alpha^2 + \alpha^2 + \dots \\
&= \frac{1}{1-\alpha} + \alpha\frac{1}{1-\alpha} + \alpha^2\frac{1}{1-\alpha} + \dots \\
&= \frac{1}{(1-\alpha)^2}
\end{aligned}
$$

Hence, the entire second sequence converges to

$$\mu\left(\sum_{i=0}^{\infty}\alpha^i(i+1) - \sum_{i=1}^{\infty}\alpha^i\right) = \left(\frac{1}{(1-\alpha)^2} - \frac{1}{1-\alpha}\right)\mu = \frac{\alpha}{(1-\alpha)^2}\mu$$

The third sequence can be decomposed as follows

$$
\begin{aligned}
& \alpha\beta(t+1) + \alpha^2\left(\beta(t+1) + \beta(t+2)\right) \\
& +\alpha^3\left(\beta(t+1) + \beta(t+2) + \beta(t+3)\right) + \dots \\
= \; & \beta\left(\alpha(t+1) + \alpha^2(2t+1+2) + \alpha^3(3t+1+2+3) + \dots\right) \\
= \; & \beta\left(\alpha t + \alpha^2 2t + \alpha^3 3t + \dots + \alpha + \alpha^2(1+2) + \alpha^3(1+2+3) + \dots\right)
\end{aligned}
$$

The first sum converges to $\frac{\alpha}{(1-\alpha)^2}t$ following the same approach as applied for the drift parameter. By rewriting the second sum, we get

$$= \beta\left(\frac{\alpha}{(1-\alpha)^2}t + \frac{\alpha}{1-\alpha} + \frac{2\alpha^2}{1-\alpha} + \frac{3\alpha^3}{1-\alpha} + \dots\right)$$

$$= \beta\left(\frac{\alpha}{(1-\alpha)^2}t + \frac{\alpha}{1-\alpha}\left(1+2\alpha+3\alpha^2+\ldots\right)\right)$$
$$= \beta\left(\frac{\alpha}{(1-\alpha)^2}t + \frac{\alpha}{(1-\alpha)}\frac{1}{(1-\alpha)^2}\right)$$
$$= \beta\frac{\alpha}{(1-\alpha)^2}t + \beta\frac{\alpha}{(1-\alpha)}\frac{1}{(1-\alpha)^2}$$

Substitution of all separately calculated components in equation (2.27) yields the desired result. Notice that a necessary condition for the convergence of all infinite sums is $|\alpha| < 1$, a restriction already derived for the forward solution above. ||

Hence, if the exogenous stochastic process x_t can be described by the random walk equation (2.26), the forward solution is a linear function depending on x_t including drift and trend component .

Consider second a *geometric random walk with drift* which can be described by

$$x_t = x_{t-1}\exp(\mu + \epsilon_t)$$

where μ stands for the drift component and ϵ_t is NIID$\sim(0,\sigma^2)\ \forall t$. Notice that by taking logs the exogenous stochastic process $\ln x_t$ follows a random walk with drift[42]. The forward looking solution is easily established and interestingly differs in its nature from the solution derived above for random walks in levels.

Proposition 2.10 If the exogenous stochastic process $\{x_t, t \in \mathcal{T}\}$ can be described as a geometric random walk with drift, equation (2.4) has the *forward solution*

$$y_t = \left(1 - \alpha\exp\left(\mu + \frac{\sigma^2}{2}\right)\right)^{-1} x_t$$

where μ denotes the drift parameter and σ^2 the variance of ϵ_t.

[42]Recall that a random walk in logs with drift is a discrete version of a geometric Brownian motion. A geometric Brownian motion can be expressed by $x_t = \exp(mt + w_t)$ where w_t denotes a simple Brownian motion with independently and normally distributed increments and μ a drift parameter. For a more extended description of geometric Brownian motions see Resnik (1992) or Karlin, Taylor (1976).

Proof of proposition 2.10 Recall[43] the forward solution of equation (2.4), i.e. $y_t = \sum_{i=0}^{\infty} \alpha^i E[x_{t+i} \mid I_t]$. By successive substitution of the geometric random walk definition, we obtain

$$y_t = x_t + \alpha E[\exp(\mu(t+1) + w_{t+1}) \mid I_t] + \alpha^2 E[\exp(\mu(t+2) + w_{t+2}) \mid I_t] + \ldots \tag{2.28}$$

Recall that $\epsilon \sim N(0, \sigma^2)$ denotes the independently, identically, and normally distributed increment of a Brownian motion. Due to independence of the increments of a Brownian motion, we can rearrange equation (2.28) by dividing the expectation term into two components such that

$$\begin{aligned} y_t &= x_t + \alpha E[\exp(\mu t + w_t) \mid I_t] E[\exp(\mu + w_{t+1} - w_t) \mid I_t] \\ &+ \alpha^2 E[\exp(\mu t + w_t) \mid I_t] E[\exp(2\mu + w_{t+2} - w_{t+1} + w_{t+1} - w_t) \mid I_t] \\ &+ \ldots \end{aligned}$$

This can equivalently be expressed in terms of the increment ϵ_t

$$\begin{aligned} y_t &= x_t + \alpha x_t E[\exp(\mu + \epsilon_{t+1}) \mid I_t] \\ &+ \alpha^2 x_t E[\exp(2\mu + \epsilon_{t+1} + \epsilon_{t+2}) \mid I_t] \\ &+ \ldots \end{aligned} \tag{2.29}$$

By using the definition of the characteristic function of normally distributed random variables and simple algebra, we can rewrite equation (2.29) as

$$\begin{aligned} y_t &= x_t + \alpha x_t \exp\left(\mu + \frac{\sigma^2}{2}\right) + \alpha^2 x_t \exp\left(2\left(\mu + \frac{\sigma^2}{2}\right)\right) + \ldots \\ &= \frac{1}{1 - \alpha \exp\left(\mu + \frac{\sigma^2}{2}\right)} x_t \end{aligned}$$

yielding the desired solution. The convergence of the infinite sum

[43] See Froot, Obstfeld (1991b) for a similar derivation of the forward solution when the exogenous process can be described as a geometric random walk in the framework of capital markets.

is guaranteed if $\alpha \exp\left(\mu + \frac{\sigma^2}{2}\right) < 1$ which can be rearranged to[44] $\mu + \frac{\sigma^2}{2} < -\ln \alpha$. ||

Notice that if the exogenous stochastic process x_t can be represented by a geometric random walk, the forward solution is proportional to the current value of the exogenous process.

2.3.1.3 Summary

In deriving a particular solution for equation (2.4), two strategies were pursued in the preceding section.

Table 2.1: Set of particular solutions

Forward solution $\lvert\alpha < 1\rvert$	**Backward solution $\lvert\alpha > 1\rvert$**
$y_t = \sum_{i=0}^{\infty} \alpha^i E[x_{t+i} \mid I_t]$	$y_t = -\sum_{i=1}^{\infty} \frac{1}{\alpha^i} x_{t-i}$
ARMA solutions {with drift}	
$\Theta(L)y_t = \frac{1}{L-\alpha}\left(L\Theta(L) - \alpha\frac{\Theta(\alpha)}{\Phi(\alpha)}\Phi(L)\right) x_t \left\{+\frac{\alpha}{1-\alpha}\frac{\Theta(\alpha)}{\Phi(\alpha)}\mu\right\}$	
AR(1) process {with drift}	
$y_t = \frac{1}{1-\alpha\phi} x_t \left\{+\frac{\alpha}{(1-\alpha)(1-\alpha\phi)}\mu\right\}$	
Random walk with drift and trend component	
$y_t = \frac{1}{1-\alpha} x_t + \frac{\alpha}{(1-\alpha)^2}\left(\mu + \frac{1}{1-\alpha}\beta\right) + \frac{\alpha}{(1-\alpha)^2}\beta t$	backward similar to forward
Geometric random walk with drift	
$y_t = \left(1 - \alpha \exp\left(\mu + \frac{\sigma^2}{2}\right)\right)^{-1} x_t$	backward similar to forward

First, we obtained a particular solution by applying forward and backward looking substitution procedures to equation (2.4) regardless of the precise nature of the stochastic process $\{x_t, t \in \mathcal{T}\}$. Second, we considered explicitly defined stochastic processes for the exogenous variable x. Two

[44] If the parameter α represents a discount factor which is often the case in an economic context, say, for instance, $\alpha = \exp(-r)$, then the condition for convergence inverts into $\mu + \frac{\sigma^2}{2} < r$. This case will be relevant in testing for bubbles below.

processes were taken into account: an ARMA–process (including a drift component), and random walk (more precisely, a random walk in levels with drift and trend parameter and a random walk in logs). Table 2.1 summarizes the resulting particular solutions.

Figure 2.2: Interdependence of the set of particular solutions

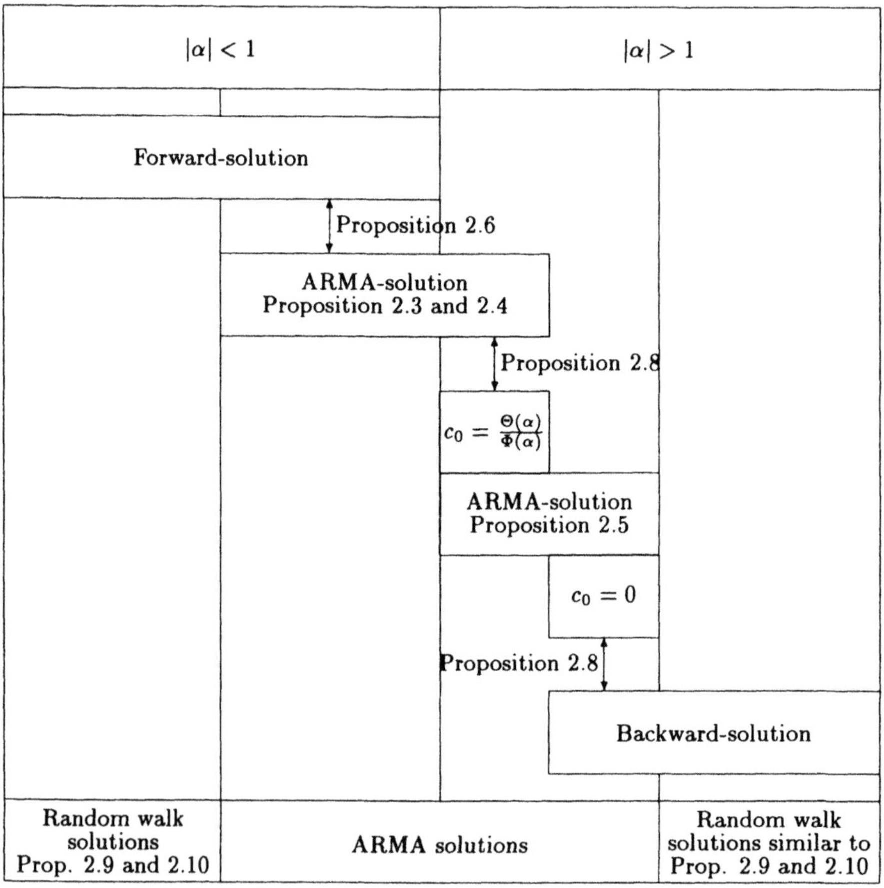

Notice that ARMA–solutions are derived regardless of the value taken by the parameter α. Figure 2.2 clarifies the interdependence between the distinct particular solutions more precisely. It turns out that in the case of an ARMA–representable exogenous process $\{x_t, t \in \mathcal{T}\}$ we must be aware of possibly overlooking particular solutions when using forward or backward solution procedures. For instance, according to proposition 2.6, the ARMA–solution is identical to the forward solution for $|\alpha| < 1$. However, according to proposition 2.8, ARMA–solutions are also available for $|\alpha| > 1$. Hence, the ARMA–solution set is broader in comparison to the solution set obtained by solely concentrating on forward or backward solutions. Additionally, by inspection of figure 2.2, the multiplicity of solutions of equation (2.4) becomes obvious. This indeterminance is mainly due to the fact that actually two endogenous variables must be solved for in equation (2.4), although only one equation is given. These two variables are the current value y_t and its expectation for the next period. The indeterminance will be further increased when the complementary solution component is considered next.

2.3.2 General solution

From the theory of deterministic difference equations, it is well known that the solution of equation (2.4) consists of a particular solution described in the preceding subsection and a complementary solution. It will be shown now that both parts are additively related and that the complementary solution equals the general solution of the homogenous counterpart of equation (2.4)[45].

Suppose the general solution of equation (2.4) can be written as

$$y_t = y_t^* + \bar{y}_t \tag{2.30}$$

where y_t^* denotes an arbitrary particular solution considered above and $\bar{y}_t$ represents an arbitrary alternative solution. Taking expectations of period $t+1$ conditional on information set I_t, we obtain

$$E[y_{t+1} \mid I_t] = E[y_{t+1}^* \mid I_t] + E[\bar{y}_{t+1} \mid I_t] \tag{2.31}$$

Substitution of equation (2.30) and (2.31) into equation (2.4) results

$$y_t^* + \bar{y}_t = \alpha E[y_{t+1}^* \mid I_t] + \alpha E[\bar{y}_{t+1} \mid I_t] + x_t$$

As we already know, the particular solution y_t^* fulfills $y_t^* = \alpha E[y_{t+1}^* \mid I_t] + x_t$ such that the alternative solution $\bar{y}_t$ must satisfy

$$\bar{y}_t = \alpha E[\bar{y}_{t+1} \mid I_t] \tag{2.32}$$

[45]See Blanchard, Fischer (1989) p 221.

which yields the desired result: the solution of the homogenous equation (2.32) can indeed be regarded as a part of the general solution of equation (2.4). In the following a martingale solution for the complementary component of equation (2.30) will be derived by following the lines described above[46]. It will turn out that bubbles most easily occur within this framework.

All stochastic processes which satisfy equation (2.32) are able to form a part of the general solution of equation (2.4). In their paper, Gourieroux, Laffont and Monfort (1982) have established the general solution of equation (2.4) as a sum of a particular and complementary solution by applying the concept of martingales. Based on the notion of conditional expectation, as considered above, we can define a martingale as follows[47].

Definition 2.5 A stochastic process $\{M_t, t \in \mathcal{T}\}$ is a *martingale* with respect to the information set I_t if it satisfies for $\forall t \in \mathcal{T}$

$$E[\mid M_t \mid] < \infty \text{ and } E[M_{t+1} \mid I_t] = M_t$$

The information set I_t contains all realizations of a vector of possibly random variables being relevant for the martingale. According to the above definition, a martingale is a stochastic process with finite expectation for any t whose value observed today M_t is the best forecast of the expected value of the process based on today's information I_t[48]. Therefore, a martingale represents a stochastic generalization of a constant process. Notice that a martingale defines a property of stochastic processes rather

[46]Recently, an alternative martingale difference approach was proposed by Broze et al. (1985) yielding often more tractable solutions, especially when deriving reduced forms of more general rational expectations models. We do not present the martingale difference approach here since the existence of rational bubbles can more easily be established in the context of the martingale solution. See also Broze, Szafarz (1991) p 38.

[47]See Taylor (1990) for a first overview of martingales. The classical reference for martingales is Doob (1953). For a more advanced treatment of martingales see among others Karlin, Taylor (1976) or any other textbook on stochastic processes. A more economic oriented approach is provided by Malliaris, Brock (1982) and Broze, Szafarz (1991).

[48]The common example for a martingale cites a gambling situation where the gambler will on average expect to be neither wealthier nor poorer than he was before the game.

than a stochastic process itself. That is the reason why the equality $E[M_{t+1} \mid I_t] = M_t$ often refers to the martingale property. Furthermore, we can define related martingale concepts which will prove useful in the remainder.

Definition 2.6 A stochastic process $\{M_t, t \in \mathcal{T}\}$ is a *submartingale (supermartingale)* with respect to the information set I_t if it satisfies for $\forall t \in \mathcal{T}$

$$E[\| M_t \|] < \infty \text{ and } E[M_{t+1} \mid I_t] \geq (\leq) M_t$$

A submartingale is a stochastic process where the (finite) expectation of a random variable always exceeds (or equals) its current value, while economic agents expect a random variable to decrease if the random variable can be characterized by a supermartingale. A submartingale (supermartingale) is a broader concept in comparison to the notion of martingales. Notice that a martingale can be described as a stochastic process which satisfies both the submartingale and the supermartingale property. Concerning the general solution of equation (2.4), the following proposition can be stated.

Proposition 2.11 The general solution of equation (2.4) can be described by the following difference equation

$$y_t = y_t^* + \frac{1}{\alpha^t} M_t$$

where y_t^* represents a particular solution of equation (2.4) and M_t is an arbitrary martingale based on the information set I_{t-1}.

Proof of proposition 2.11 See Gourieroux, Laffont and Monfort (1982). Consider the homogenous part of equation (2.4)

$$\bar{y}_t = \alpha E[\bar{y}_{t+1} \mid I_t] \tag{2.33}$$

Recall that any complementary solution has to satisfy equation (2.33). Multiply both sides of equation (2.33) by α^t

$$\alpha^t \bar{y}_t = \alpha^{t+1} E[\bar{y}_{t+1} \mid I_t] \tag{2.34}$$

and define a stochastic process $\{M_t, t \in \mathcal{T}\}$ as

$$M_t = \alpha^t \bar{y}_t \tag{2.35}$$

where $\bar{y}$ satisfies (2.33). The stochastic process $\{M_t, t \in \mathcal{T}\}$ is a martingale since by using equation (2.33)

$$E[M_{t+1} \mid I_t] = E[\alpha^{t+1}\bar{y}_{t+1} \mid I_t] = \alpha^{t+1}E[\bar{y}_{t+1} \mid I_t] = \alpha^t \bar{y}_t = M_t$$

Notice that the definitorial equation (2.35) satisfies equation (2.33). Therefore, the general solution of the homogenous equation is easily obtained by rearranging equation (2.35)

$$\bar{y}_t = \frac{1}{\alpha^t} M_t$$

where M_t stands for an arbitrary martingale process. ||

Summarizing, the general solution of equation (2.4) can be described as the sum of a particular solution and a complementary solution (general solution of the homogenous counterpart of the original equation), which corresponds to the martingale solution and can be expressed as

$$y_t = y_t^* + \frac{1}{\alpha^t} M_t$$

where y_t^* denotes the particular solution and M_t an arbitrary martingale. Notice that the occurrence of a martingale $\{M_t, t \in \mathcal{T}\}$ within the general solution of equation (2.4) increases the indeterminance of the general solution already established in the context of the particular solution. Any stochastic process satisfying the martingale property can be considered as a component of the general solution. In many economic applications both components of the general solution have been given special names.

Definition 2.7 The general solution of equation (2.4) can be divided into two components

$$y_t = f_t + b_t$$

whereas $f_t = y_t^*$ denotes the *fundamental solution* usually characterizing the long–run equilibrium of the endogenous stochastic process (it corresponds to the particular solution[49]) and

$$b_t = \frac{1}{\alpha^t} M_t \qquad (2.36)$$

is called the *bubble solution* characterizing the (permanent or temporary) deviation of the endogenous stochastic process from the long–run equilibrium. Since the bubble solution appears within a rational expectations framework, this solution component is also–called a rational bubble.

Concerning the relationship between martingales and bubbles in equation (2.36), we can state the following proposition.

Proposition 2.12 The bubble process $\{b_t, t \in \mathcal{T}\}$ of equation (2.36) is either a submartingale or a supermartingale depending on the value of the parameter $\alpha > 0$, i.e. for $\alpha \leq 1$, b_t follows a submartingale and for $\alpha \geq 1$, b_t follows a supermartingale.

Proof of proposition 2.12 Rearranging equation (2.36) as $b_t = \alpha^{-t} M_t$, leading once and taking expectations on both sides we obtain

$$E[b_{t+1} \mid I_t] = E[\alpha^{-(t+1)} M_{t+1} \mid I_t] = \alpha^{-(t+1)} E[M_{t+1} \mid I_t]$$

Using the definition of martingales and the relationship between b_t and M_t defined in equation (2.36) we get

$$\alpha^{-(t+1)} E[M_{t+1} \mid I_t] = \alpha^{-(t+1)} M_t = \alpha^{-(t+1)} \alpha^t b_t = \frac{1}{\alpha} b_t$$

Hence, we can conclude that the bubble solution obeys

$$E[b_{t+1} \mid I_t] = \frac{1}{\alpha} b_t \qquad (2.37)$$

and, therefore,

1. $E[b_{t+1} \mid I_t] \geq b_t$ for $\alpha \leq 1$
2. $E[b_{t+1} \mid I_t] \leq b_t$ for $\alpha \geq 1$

[49]Notice that some authors consider only the forward solution as the fundamental solution. See Broze, Szafarz (1991) p 31, for instance.

establishing the desired properties of (sub/super)martingales. ||

This proposition clarifies the relation between bubbles and martingales often misdescribed in literature: a bubble itself usually does not follow a martingale, it only contains a martingale as a component. Depending on the value taken by the parameter α a bubble can be described by a sub– or a supermartingale.

2.4 Bubble processes

So far, a particular and a complementary component characterizing the fundamental and bubble solution of equation (2.4) have been derived. In economic theory, however, the complementary solution was neglected for a long time. The endogenous stochastic process was exclusively described by the fundamental solution by imposing partly unreasonable constraints which rule out bubbles[50]. But self–restricting to the fundamental solution proved insufficient in describing reality. The derivability of bubble solutions gave rise to the hope for a better explanation of economic phenomena by incorporating an additional solution component which enables the endogenous stochastic process to deviate temporarily, or permanently, from its fundamentals in the sense that the endogenous stochastic process can lie above its fundamentals[51] – a phenomenon often observed in the data.

Due to the concept of martingales occurring in the general solution of equation (2.4), different bubble processes may be distinguished. Despite the obvious generality of martingales, only a few bubble processes have been considered in literature so far. The aim of this section is to increase the set of rational bubbles by providing its common origin in the form of a rather general martingale process. More concrete martingales and their bubble counterparts can easily be derived from this common root by adopting special assumptions concerning the underlying parameters. The resulting set of bubble processes include both known and unknown bubble

[50]See, for instance, Taylor (1977) and Gourieroux et al. (1982).

[51]Notice that on economic grounds only positive bubble processes are reasonable. For a further treatment of this issue see section 2.4.4.

specifications. Hence, a procedure based on this common root provides a systematic way for deriving bubble processes.

By taking advantage of the idea that every random walk is a martingale, this general martingale process can be considered as a generalization of a stochastic unit root (STUR) process first proposed by Granger, Swanson (1994). By introducing an additional stochastic coefficient ζ_t for the error term, we obtain a generalized stochastic martingale process (GSTM) M_t, i.e.

$$M_t = \rho_t M_{t-1} + \zeta_t \epsilon_t \tag{2.38}$$

where ϵ_t is an arbitrary error term satisfying $E[\epsilon_{t+j} \mid I_t] = 0, \ \forall j > 0$ with respect to $I_t = \{\epsilon_t, \epsilon_{t-1}, \ldots\}$. The random variable ρ_t is supposed to have a conditional expectation equal to 1, i.e. $E[\rho_t \mid I_{t-1}] = 1, \ \forall t > 0$. Concerning the stochastic nature of the random variable ζ_t, no restricting assumption is necessary. The correlation structure between the random variables can be summarized as follows

$$\begin{aligned} &E[\rho_t M_t \mid I_t] = 0, \ \ E[\rho_t \epsilon_t \mid I_t] = 0, \ \ E[\rho_t \zeta_t \mid I_t] = 0 \\ &E[\zeta_t M_t \mid I_t] = 0, \ \ E[\zeta_t \epsilon_t \mid I_t] = 0, \ \ E[\epsilon_t M_t \mid I_t] = 0 \end{aligned}$$

establishing the independence of all stochastic variables in all leads and lags. It can easily be shown that the generalized stochastic martingale process (2.38) indeed represents a martingale since

$$E[M_{t+1} \mid I_t] = E[\rho_{t+1} \mid I_t] E[M_t \mid I_t] + E[\zeta_{t+1} \mid I_t] E[\epsilon_{t+1} \mid I_t] = M_t$$

Two different cases can be regarded. First, a STUR–martingale similar to Granger, Swanson (1994) STUR–specification is considered by setting ζ_t constant and equal to 1. It becomes clear that most of the bubble processes can be derived by using a STUR–martingale. Then, by examining the case $\rho_t = 1$, it will turn out that the set of a special kind of rational bubbles, the so–called intrinsic bubbles, can be extended.

CASE 1: $\zeta_t = 1$

Starting with a STUR–martingale, i.e.

$$M_t = \rho_t M_{t-1} + \epsilon_t \tag{2.39}$$

with $E[\epsilon_{t+j} \mid I_t] = 0, \forall j > 0$ and $E[\rho_t \mid I_{t-1}] = 1, \forall t > 0$ we are interested in the special nature of the random variable ρ_t. Suppose a random variable $\nu_t \sim N(\mu, \sigma_\nu^2)$. For an arbitrary λ it can be shown that for $\rho_t = \exp(\lambda\nu_t - (\lambda\mu + \frac{1}{2}\lambda^2\sigma_\nu^2))$ the assumption $E[\rho_{t+1} \mid I_t] = 1$ is satisfied due to the characteristic function of normally distributed random variables, i.e. since $E[\exp(\lambda\nu_t)] = \exp(\lambda\mu + \frac{1}{2}\lambda^2\sigma_\nu^2)$, $E[\rho_t] = 1$ is satisfied. Substituting this characterization of ρ_t into equation (2.39) the following martingale process is obtained

$$M_t = \exp\left(\lambda\nu_t - \left(\lambda\mu + \frac{1}{2}\lambda^2\sigma_\nu^2\right)\right) M_{t-1} + \epsilon_t \tag{2.40}$$

Several special cases can be distinguished for different parameter combinations, in particular for different values of the distributional parameters of ν_t and the arbitrary parameter λ. These parameter specifications generally try to simplify the exponential expression in equation (2.40). The corresponding martingales are summarized in table 2.2.

Notice that we distinguish between $\sigma_\epsilon^2 = 0$ and $\sigma_\epsilon^2 \neq 0$. This can be justified by the possibility of deriving different bubble processes whether ϵ_t is stochastic or not. Based on these martingales derived from a general STUR–martingale, the corresponding basic bubble processes can easily be established. As already shown, bubble processes are related to martingales through[52]

$$b_t = \frac{1}{\alpha^t} M_t \tag{2.41}$$

In order to obtain the corresponding bubble process, consider now the STUR–martingale, i.e. $M_t = \rho_t M_{t-1} + \epsilon_t$ rearranged by using the common lag operator L

$$M_t = \frac{1}{1 - \rho_t L}\epsilon_t \tag{2.42}$$

Substitution of equation (2.42) into equation (2.41) and simple rearrange-

---

[52]Besides relation (2.41) obtained in the context of solving for the general solution of equation (2.4), alternative relations may be useful. For instance, equation (2.41) can equivalently be expressed in exponential terms $b_t = \exp(-t \ln \alpha) M_t$. In some situations the ratio of martingales of two successive periods becomes relevant. From equation (2.41) we get $\frac{b_t}{b_{t-1}} = \frac{1}{\alpha}\frac{M_t}{M_{t-1}}$ or equivalently expressed in exponential terms $\frac{b_t}{b_{t-1}} = \exp(-\ln \alpha)\frac{M_t}{M_{t-1}}$.

Table 2.2: Set of martingales

	$\sigma_\epsilon^2 = 0$
arbitrary μ, σ_ν^2 arbitrary λ	M1: General stochastic martingale $M_t = \exp(\lambda\nu_t - (\lambda\mu + \frac{1}{2}\lambda^2\sigma_\nu^2))M_{t-1}$ with $\nu_t \sim N(\mu, \sigma_\nu^2)$
arbitrary μ, σ_ν^2 $\lambda = 0$	M2: Constant martingale $M_t = M_{t-1} = b,\ b > 0$
arbitrary μ, σ_ν^2 $\lambda = -\frac{2\mu}{\sigma_\nu^2}$	M3: Geometric martingale $M_t = \exp(-\frac{2\mu}{\sigma_\nu^2}\nu_t)M_{t-1}$ with $\nu_t \sim N(\mu, \sigma_\nu^2)$
$\mu = -\frac{\sigma_\nu^2}{2}$ $\lambda = -\frac{2\mu}{\sigma_\nu^2} \Rightarrow \lambda = 1$	M3a: Restricted geometric martingale $M_t = \exp(\nu_t)M_{t-1}$ with $\nu_t \sim N(-\frac{\sigma_\nu^2}{2}, \sigma_\nu^2)$
	$\sigma_\epsilon^2 \neq 0$
arbitrary μ, σ_ν^2 arbitrary λ	M5: General stochastic martingale incl. ϵ_t $M_t = \exp(\lambda\nu_t - (\lambda\mu + \frac{1}{2}\lambda^2\sigma_\nu^2))M_{t-1} + \epsilon_t$ with $\nu_t \sim N(\mu, \sigma_\nu^2)$
arbitrary μ, σ_ν^2 $\lambda = 0$	M6: Random walk martingale $M_t = M_{t-1} + \epsilon_t$
arbitrary μ, σ_ν^2 $\lambda = -\frac{2\mu}{\sigma_\nu^2}$	M7: Geometric martingale incl. ϵ_t $M_t = \exp(-\frac{2\mu}{\sigma_\nu^2}\nu_t)M_{t-1} + \epsilon_t$ with $\nu_t \sim N(\mu, \sigma_\nu^2)$
$\mu = -\frac{\sigma_\nu^2}{2}$ $\lambda = -\frac{2\mu}{\sigma_\nu^2} \Rightarrow \lambda = 1$	M7a: Restricted geometric martingale incl. ϵ_t $M_t = \exp(\nu_t)M_{t-1} + \epsilon_t$ with $\nu_t \sim N(-\frac{\sigma_\nu^2}{2}, \sigma_\nu^2)$

ment yields

$$\begin{aligned} b_t &= \frac{1}{(1-\rho_t L)\alpha^t}\epsilon_t \\ (1-\rho_t L)\alpha^t b_t &= \epsilon_t \\ \alpha^t b_t &= \rho_t \alpha^{t-1} b_{t-1} + \epsilon_t \\ b_t &= \frac{\rho_t}{\alpha} b_{t-1} + u_t \quad \text{with} \quad u_t = \frac{1}{\alpha^t}\epsilon_t \end{aligned} \tag{2.43}$$

For the different specifications of the random variable ρ_t in table 2.2 the corresponding basic bubble processes can analogously be derived by applying relation (2.43) or equivalently defined relations. The resulting bubble processes are tabulated in table 2.3. Notice that u_t satisfies $E[u_{t+j} \mid I_t] = 0$, $\forall j > 0$.

Table 2.3: Set of basic bubble processes

	$\sigma_\epsilon^2 = 0$
arbitrary μ, σ_ν^2 arbitrary λ	B1: General stochastic basic bubble $b_t = \exp(\lambda\nu_t - (\lambda\mu + \frac{1}{2}\lambda^2\sigma_\nu^2 + \ln\alpha))b_{t-1}$ with $\nu_t \sim N(\mu, \sigma_\nu^2)$
arbitrary μ, σ_ν^2 $\lambda = 0$	B2: Deterministic bubble $b_t = \frac{1}{\alpha^t}b,\ b > 0$
arbitrary μ, σ_ν^2 $\lambda = -\frac{2\mu}{\sigma_\nu^2}$	B3: Geometric basic bubble I $b_t = \frac{1}{\alpha}\exp(\nu_t)^{-\frac{2\mu}{\sigma_\nu^2}}b_{t-1}$ with $\nu_t \sim N(\mu, \sigma_\nu^2)$
$\mu = -\frac{\sigma_\nu^2}{2}$ $\lambda = -\frac{2\mu}{\sigma_\nu^2} \Rightarrow \lambda = 1$	B3a: Restricted geometric basic bubble I $b_t = \frac{1}{\alpha}\exp(\nu_t)b_{t-1}$ with $\nu_t \sim N(-\frac{\sigma_\nu^2}{2}, \sigma_\nu^2)$
arbitrary μ, σ_ν^2 λ s.t. $\lambda\mu + \frac{1}{2}\lambda^2\sigma_\nu^2 + \ln\alpha = 0$	B4: Geometric basic bubble II $b_t = \exp(\nu_t)^\lambda b_{t-1}$ with $\nu_t \sim N(\mu, \sigma_\nu^2)$
	$\sigma_\epsilon^2 \neq 0$
arbitrary μ, σ_ν^2 arbitrary λ	B5: General stochastic basic bubble incl. ϵ_t $b_t = \exp(\lambda\nu_t - (\lambda\mu + \frac{1}{2}\lambda^2\sigma_\nu^2 + \ln\alpha))b_{t-1} + u_t$ with $\nu_t \sim N(\mu, \sigma_\nu^2)$
arbitrary μ, σ_ν^2 $\lambda = 0$	B6: Near random walk bubble $b_t = \frac{1}{\alpha}b_{t-1} + u_t$
arbitrary μ, σ_ν^2 $\lambda = -\frac{2\mu}{\sigma_\nu^2}$	B7: Geometric basic bubble I incl. ϵ_t $b_t = \frac{1}{\alpha}\exp(\nu_t)^{-\frac{2\mu}{\sigma_\nu^2}}b_{t-1} + u_t$ with $\nu_t \sim N(\mu, \sigma_\nu^2)$
$\mu = -\frac{\sigma_\nu^2}{2}$ $\lambda = -\frac{2\mu}{\sigma_\nu^2} \Rightarrow \lambda = 1$	B7a: Restricted geometric basic bubble I incl. ϵ_t $b_t = \frac{1}{\alpha}\exp(\nu_t)b_{t-1} + u_t$ with $\nu_t \sim N(-\frac{\sigma_\nu^2}{2}, \sigma_\nu^2)$
arbitrary μ, σ_ν^2 λ s.t. $\lambda\mu + \frac{1}{2}\lambda^2\sigma_\nu^2 + \ln\alpha = 0$	B8: Geometric basic bubble II incl. ϵ_t $b_t = \exp(\nu_t)^\lambda b_{t-1} + u_t$ with $\nu_t \sim N(\mu, \sigma_\nu^2)$

Some common characteristics of the resulting basic bubble processes can be pointed out:

- Both sets of bubbles B1-B4 and B5–B8 are conceptually equal. The difference between both lies in the additional source of randomness introduced by the variance of the error term ϵ_t of the STUR–martingale assumed to be non–zero in the latter subset.

- The most general stochastic basic bubble specification is represented by B1 (B5) since no restrictions concerning the random coefficient ρ_t are imposed. B1 (B5) can simply be obtained by taking the general stochastic martingale M1 (M5) and by applying the relation described in equation (2.43) for $\rho_t = \exp(\lambda\nu_t - (\lambda\mu + \frac{1}{2}\lambda^2\sigma_\nu^2))$. We get

$$(2.44) \qquad b_t = \exp\left(\lambda\nu_t - \left(\lambda\mu + \frac{1}{2}\lambda^2\sigma_\nu^2 + \ln\alpha\right)\right) b_{t-1} \quad \text{with } \nu_t \sim N(\mu, \sigma_\nu^2)$$

The subsequent bubble specifications B2–B4 (B6–B8) mainly try to simplify the exponential expression in equation (2.44) and, therefore, represent subcases of the general stochastic bubble process B1 (B5). For instance,

$$b_t = \exp(\underbrace{\lambda\nu_t - (\lambda\mu + \frac{1}{2}\lambda^2\sigma_\nu^2}_{=0 \text{ for B2 (B6)}} + \ln\alpha))b_{t-1}$$

$$b_t = \exp(\lambda\nu_t - \underbrace{(\lambda\mu + \frac{1}{2}\lambda^2\sigma_\nu^2}_{=0 \text{ for B3/3a (B7/7a)}} + \ln\alpha))b_{t-1}$$

$$b_t = \exp(\lambda\nu_t - \underbrace{(\lambda\mu + \frac{1}{2}\lambda^2\sigma_\nu^2 + \ln\alpha)}_{=0 \text{ for B4 (B8)}})b_{t-1}$$

Notice that the last bubble specification B4 (B8) yields two possible values for the parameter λ related to the solution of the corresponding quadratic equation[53].

- Based on the above considerations, it becomes obvious that B3a (B7a) represents a subcase of B3 (B7) by additionally restricting the relation between the parameters μ and σ_ν^2 after already having predetermined the parameter λ[54].

[53] The quadratic equation will be discussed more precisely in the context of intrinsic bubbles since it has a greater relevance there.

[54] A bubble specification similar to B3a (B7a) was recently proposed by Charemza, Deadman (1995). See section 4.1.5 for a more detailed treatment of the Charemza, Deadman (1995) approach.

- All bubbles are stochastic either through the random coefficient ρ_t or through an additional error term u_t except of B2 representing a deterministic bubble.

- Except for B2 and B6, all bubble processes contain the random variable ν_t. This will be important later when introducing new bubble processes depending on other variables rather than on lagged values of its own.

- The special bubble processes B2–B4 (B6–B8) in table 2.3 are only a small selection of theoretically derivable bubble processes. Notice that any process satisfying the general bubble process description B1 (B5) for an arbitrary parameter λ can be regarded as a bubble process.

Based on the results summarized in table 2.3 and on additional considerations undertaken on the subsequent pages, the following general classes of bubble processes may be distinguished:

1. A bubble depending on its own observation of the previous period is called a *Markovian bubble*[55].

$$b_t = g(b_{t-1})$$

2. A bubble depending on the exogenous stochastic (i.e. fundamental) process $\{x_t, t \in \mathcal{T}\}$ is called an *intrinsic bubble*[56].

$$b_t = g(x_t)$$

3. A bubble depending on an arbitrary exogenous, possibly stochastic, process $\{z_t, t \in \mathcal{T}\}$ or any other event occurring outside the model is called an *extrinsic bubble*[57].

$$b_t = g(z_t)$$

These bubble classes will now be examined in greater detail.

---

[55]Some authors call this type of bubble time–dependent (e.g. Ikeda, Shibata (1992)) or simply growing bubbles (e.g. Camerer (1989)).

[56]The notion of intrinsic bubbles is due to Froot and Obstfeld (1991b).

[57]This type of bubble is called extraneous bubble, as well.

2.4.1 Markovian bubbles

A Markovian bubble can be characterized as depending only on its own realization of the preceding period like Markov processes usually do. It turns out that table 2.3, summarizing the basic bubble processes, contains two theoretically derivable Markovian bubble processes, i.e. the deterministic bubble B2 and the near random walk bubble B6[58].

2.4.1.1 Deterministic bubble

The simplest characterization of a Markovian bubble is the deterministic bubble process B2 in table 2.3. Recall that the deterministic bubble B2 is based on the martingale M2, i.e. $M_t = M_{t-1}$ corresponding to a steady process $M_t = b$ where $b > 0$[59] is an arbitrary constant. The deterministic bubble can easily be derived by simply applying relation (2.41)[60]. The following specification is obtained

$$b_t = \frac{1}{\alpha^t} b$$

Concerning the behaviour of a deterministic bubble, two cases have to be distinguished: For $|\alpha| < 1$ the bubble process $\{b_t, t \in \mathcal{T}\}$ grows monotonically for $0 < \alpha < 1$ or oscillates for $-1 < \alpha < 0$ without end at rate $\frac{1}{|\alpha|}$ and is, therefore, usually called an 'ever expanding bubble'. Its potential for explaining real world phenomena is limited since usually we do not observe economic variables growing without end, at least if we assume a permanent presence of an ever expanding bubble. This bubble specification, however, may have explanatory power in the case of a temporary presence of rational bubbles in time series[61].

---

[58]Notice that all remaining bubble processes in table 2.3 include an additional random variable ν_t which exclude these processes from the set of Markovian bubbles. However, it will turn out that the random variable ν_t becomes relevant in the context of alternative bubble classes.

[59]The case $b < 0$ is at least economically not reasonable.

[60]Equivalently, we can take the general stochastic basic bubble process B1 and set the parameter λ equal to zero.

[61]A context in which considering deterministic bubbles seems to be reasonable is a period of hyperinflation. See Flood, Garber (1980) for a treatment of deterministic bubbles in the context of German hyperinflation during the 1930s.

Contrarily, for $|\alpha| > 1$ the bubble process approaches zero, monotonically for $\alpha > 1$ or oscillating for $\alpha < -1$. A bubble specification of this kind has no explanatory power in the long run since the bubble vanishes as time goes by. That is the reason why deterministic bubbles are usually neglected in the case $|\alpha| > 1$.

Summarizing, a deterministic bubble appears to be applicable if ever expanding phenomena are observable in the data, usually only temporarily, bearing in mind that the parameter $|\alpha|$ ought to be smaller than 1. While deterministic bubbles do not contain any random component, all subsequent bubble processes can be considered as stochastic.

2.4.1.2 Near random walk bubble

The second Markovian bubble directly derivable from table 2.2 is a near random walk bubble. Recall that a near random walk bubble is related to the near random walk martingale specification M6, i.e. $M_t = M_{t-1} + \epsilon_t$ where $E[\epsilon_{t+j} \mid I_t] = 0, \forall j > 0$[62]. For deriving the corresponding near random walk bubble, the martingale M6 can be transformed into $M_t = \frac{1}{1-L}\epsilon_t$ by using lag operator L yielding the moving-average representation of a random walk, which consists of an infinite sum of white noises. By using relation (2.41), the corresponding near random walk bubble specification is easily established

$$b_t = \frac{1}{\alpha} b_{t-1} + u_t \quad \text{with} \quad u_t = \frac{1}{\alpha^t}\epsilon_t$$

The resulting bubble process looks like an AR(1)–process for $|\alpha| > 1$. For $|\alpha| < 1$ it shows explosive behaviour. Unfortunately, an economic application of near random walk bubbles has not been found yet. Both the theoretical meaning of such a bubble specification and the empirical relevance are still uncertain. However, it laid the basis for the development of the next Markovian bubble specification.

[62] Notice that a random walk requiress ϵ_t to be white noise. The restriction concerning ϵ_t applied here is weaker.

2.4.1.3 Bursting bubble

Besides these two special Markovian bubble processes directly gained from table 2.3, there exists another bubble process, a so-called ***bursting bubble*** which can be derived only indirectly. Bursting bubbles allow for describing phenomena like crashes, where the notion of 'bubbles' originally comes from[63], and can easily be constructed: Extend the STUR-martingale (2.39) simply by considering the following transition probability matrix

$$(2.45) \qquad M_t = \begin{cases} \frac{\rho_t}{\pi} M_{t-1} + \epsilon_t & \text{with probability } \pi \\ \epsilon_t & \text{with probability } 1 - \pi \end{cases}$$

where $\{\epsilon_t, t \in \mathcal{T}\}$ is $E[\epsilon_{t+j} \mid I_t] = 0$, $\forall j > 0$ and $0 \leq \pi \leq 1$. The stochastic process $\{M_t, t \in \mathcal{T}\}$ defined in equation (2.45) is indeed a martingale with respect to I_t with $I_t = \{\epsilon_t, \epsilon_{t-1}, \ldots\}$ since

$$\begin{aligned} E[M_{t+1} \mid I_t] &= E\left[\pi\left(\frac{\rho_{t+1}}{\pi} M_t + \epsilon_{t+1}\right) + (1-\pi)\epsilon_{t+1} \mid I_t\right] \\ &= E[\rho_{t+1} M_t + \epsilon_{t+1} \mid I_t] = M_t \end{aligned}$$

Due to its probabilistic nature, the process (2.45) can be viewed as a bursting martingale. By substituting the bursting martingale (2.45) into relation (2.25), the corresponding bursting bubble process is obtained.

$$(2.46) \qquad b_t = \begin{cases} \frac{\rho_t}{\pi\alpha} b_{t-1} + u_t & \text{with probability } \pi \\ u_t & \text{with probability } 1 - \pi \end{cases}$$

with $0 \leq \pi \leq 1$[64]. For every bubble process tabulated in table 2.3 and summarized in the general bubble description (2.43) there exists a transition probability matrix which leads to a bursting bubble specification. In particular, the most often cited bursting bubble is the specification based on the near random walk bubble discussed above, i.e.

$$b_t = \begin{cases} \frac{1}{\pi\alpha} b_{t-1} + u_t & \text{with probability } \pi \\ u_t & \text{with probability } 1 - \pi \end{cases}$$

---

[63]The notion of 'bursting bubbles' is originally due to Blanchard (1979b). See also Blanchard and Watson (1982).

[64]Notice that a bursting bubble represents a generalization of the above mentioned near random walk bubble which is obtained by taking $\pi = 1$.

It is easily seen that this bursting bubble grows at rate $\frac{1}{\pi\alpha}$ with probability π and bursts in every period with probability $1 - \pi$. Notice that the growth rate $\frac{1}{\pi\alpha}$ depends on the probability of further bubble growth π. Since $\pi \leq 1$, a bursting bubble grows temporarily faster in comparison to a deterministic bubble increasing only at rate $\frac{1}{\alpha}$. This is due to the fact that the risk of bursting is compensated by a higher growth rate. Only the expected growth rate equals the growth rate of the deterministic bubble[65]. Simultaneously bursting bubbles avoid the unappealing unlimited–growth property of deterministic bubbles since they terminate with probability one.

To conclude, so far we have considered three different Markovian bubble processes: a deterministic, a near random walk and a bursting bubble. Now we turn to the remaining bubble processes of table 2.3 containing the random variable ν_t, which forms the basis for intrinsic and extrinsic bubbles.

2.4.2 Intrinsic bubbles

The idea of intrinsic bubbles is to relate both the bubble and the fundamental solution by assuming that the bubble process itself depends exclusively on the fundamental process. Usually, the bubble process is supposed to react positively to changes in the fundamental process in the sense that if the fundamental process increases, the bubble process should do the same. This characterization of intrinsic bubbles provides a reasonable description of an overreaction to changes in fundamentals–effect observed in the data[66].

---

[65] An extension of the concept of bursting bubbles comes from the consideration that the probability of further bubble growth π may well be a function of how long the bubble has lasted, or of how far the price is from its fundamentals. Thus, it seems to be reasonable to assume a distribution function π_b depending on the bubble process $\{b_t, t \in \mathcal{T}\}$ with $\pi_b = F(b_t)$ where F denotes any arbitrary distribution function. Notice that now the growth rate itself $\frac{1}{\pi_b\alpha}$ increases when the bubble becomes bigger and bigger. Consequently, this leads to a growth rate temporarily higher than in the case of constant probability to burst. See Blanchard, Watson (1982).

[66] See, section 3.1.2 for a further description.

Consequently, in order to derive intrinsic bubbles the underlying fundamental stochastic process $\{x_t, t \in \mathcal{T}\}$ has to be specified. Two fundamental processes will be examined here: a non-stationary random walk and a stationary ARMA-process. Assuming fundamentals follow a random walk, table 2.3 is useful for deriving the set of possible intrinsic bubble specifications. Assuming fundamentals follow an ARMA-process, a similar approach like that applied until now will be adopted based on considerations concerning the GSTM process $M_t = \rho_t M_{t-1} + \zeta_t \epsilon_t$ introduced above. Additionally, an alternative procedure based on the method of undetermined coefficients is presented for the latter case which will prove easier to apply.

2.4.2.1 Fundamentals as a random walk

The most apparent intrinsic bubble specification concerns fundamentals describable as a (pure) random walk, i.e.

$$x_t = x_{t-1} + \epsilon_t$$

where ϵ_t is usually assumed to be white noise[67]. Since every random walk is a martingale, the corresponding bubble is easily established without using the approach presented above. The bubble solution can be stated as

$$b_t = \frac{1}{\alpha^t} x_t \tag{2.47}$$

The general procedure for deriving intrinsic bubbles for more general random walk processes is to substitute appropriate random variables for $\nu_t \sim N(\mu, \sigma_\nu^2)$ incorporated in the basic bubble processes presented above. Hence, based on the results of table 2.3, an instrinsic bubble process can be found for every bubble process containing ν_t.

Suppose the fundamental process follows a *random walk with drift* μ, i.e.

$$x_t - x_{t-1} = \mu + \eta_t \tag{2.48}$$

where $\eta_t \sim N(0, \sigma_\nu^2)$. Substitution of expression (2.48) for ν_t into the general stochastic basic bubble characterization B1 yields

$$b_t = \exp(\lambda(x_t - x_{t-1}) - (\lambda\mu + \frac{1}{2}\lambda^2\sigma_\nu^2 + \ln\alpha))b_{t-1}$$

[67]Recall that $E[\epsilon_{t+1} \mid I_t] = 0$ is sufficient.

By extending the second term of the exponential expression, we can derive

$$\begin{aligned} \frac{b_t}{b_{t-1}} &= \frac{\exp(\lambda x_t - (\lambda\mu + \frac{1}{2}\lambda^2\sigma_\nu^2 + \ln\alpha)t)}{\exp(\lambda x_{t-1} - (\lambda\mu + \frac{1}{2}\lambda^2\sigma_\nu^2 + \ln\alpha)(t-1))} \\ \Rightarrow \quad b_t &= \exp(\lambda x_t - (\lambda\mu + \frac{1}{2}\lambda^2\sigma_\nu^2 + \ln\alpha)t) \end{aligned}$$

Corresponding to different values taken by λ[68], table 2.4 shows the resulting intrinsic bubble processes if fundamentals follow a random walk with drift.

Table 2.4: Set of intrinsic bubble processes (random walk with drift)

arbitrary μ, σ_ν^2 arbitrary λ	B1 $\Rightarrow$ I1 General intrinsic bubble $b_t = \exp(\lambda\mu + \frac{1}{2}\lambda^2\sigma_\nu^2 + \ln\alpha)^{-t}\exp(x_t)^\lambda$
arbitrary μ, σ_ν^2 $\lambda = 0$	B2 $\Leftrightarrow$ I2 Deterministic bubble $b_t = \frac{1}{\alpha^t}b$, $b = 1$
arbitrary μ, σ_ν^2 $\lambda = -\frac{2\mu}{\sigma_\nu^2}$	B3 $\Rightarrow$ I3: Geometric intrinsic bubble I $b_t = \frac{1}{\alpha^t}\exp(x_t)^{-\frac{2\mu}{\sigma_\nu^2}}$
$\mu = -\frac{\sigma_\nu^2}{2}$ $\lambda = -\frac{2\mu}{\sigma_\nu^2} \Rightarrow \lambda = 1$	B3a $\Rightarrow$ I3a: Restricted geometric intrinsic bubble I $b_t = \frac{1}{\alpha^t}\exp(x_t)$
arbitrary μ, σ_ν^2 λ s.t. $\lambda\mu + \frac{1}{2}\lambda^2\sigma_\nu^2 + \ln\alpha = 0$	B4 $\Rightarrow$ I4: Geometric intrinsic bubble II $b_t = \exp(x_t)^{\lambda_{1,2}}$

The following remarks concerning intrinsic bubbles tabulated in table 2.4 can be made:

- Since the presence of an error term u_t makes the derivation of intrinsic bubbles impossible, the intrinsic bubble specifications introduced here relate only to the case $\sigma_\epsilon^2 = 0$.

- I1 related to B1 represents the most general intrinsic bubble process. The subceeding intrinsic bubble processes can be considered as subcases of I1 simultaneously corresponding to their basic bubble counterparts in table 2.3. Also the subcase property of I3a related to I3 remains valid.

[68] Alternatively, the fundamental process (2.48) can be substituted directly into the bubble specifications of table 2.3.

- In the case of geometric intrinsic bubble I4 two roots are usually obtained corresponding to the characteristic equation $\lambda\mu + \frac{1}{2}\lambda^2\sigma_\nu^2 + \ln\alpha = 0$. Notice that the two roots are only defined for $\alpha > 0$. It is easily seen that both roots are real only if $\mu^2 > \sigma^2 \ln\alpha$. Moreover, it often turns out that we can restrict the value of α being smaller than one on economic grounds, which guarantees that we obtain two real roots, one positive and one negative.

- All instrinsic bubbles are stochastic except for B10 representing the same deterministic bubble as B2. Notice that the constant b is restricted to be equal to one here.

- Any fundamental process x_t following a random walk with drift and satisfying the general bubble description in table 2.4 in connection with any arbitrary value for λ can be regarded as an intrinsic bubble. Hence, the special intrinsic bubbles noted in table 2.4 are by no means exclusive.

Finally, it should be emphasized that through including an arbitrary parameter λ the problem of multiple solutions has increased again. By assuming fundamentals follow a random walk, the corresponding intrinsic bubble processes are multiple in the sense that every specification theoretically satisfying the most general intrinsic bubble description B9 can be applied for proving the existence of intrinsic bubbles. Notice further that only the geometric intrinsic bubble II process I4 has been investigated in literature so far[69].

By applying the same procedure as before, the set of intrinsic bubbles in the case of fundamentals following a *geometric random walk* can easily be established. It will turn out that the bubble processes are very similar to those of the solution set of table 2.4. Suppose the fundamental process can be described as a geometric random walk being equal to a random walk in logs including a drift component μ, i.e.

$$\ln x_t - \ln x_{t-1} = \mu + \eta_t \tag{2.49}$$

[69]See Froot, Obstfeld (1991b).

where $\eta_t \sim N(0, \sigma_\nu^2)$. Substituting expression (2.49) for ν_t in the general stochastic basic bubble characterization B1 yields

$$b_t = \exp(\lambda(\ln x_t - \ln x_{t-1}) - (\lambda\mu + \frac{1}{2}\lambda^2\sigma_\nu^2 + \ln\alpha))b_{t-1}$$

Again, by extending the second term in the exponential expression, we can derive

$$\begin{aligned} \frac{b_t}{b_{t-1}} &= \frac{\exp(\lambda \ln x_t - (\lambda\mu + \frac{1}{2}\lambda^2\sigma_\nu^2 + \ln\alpha)t)}{\exp(\lambda \ln x_{t-1} - (\lambda\mu + \frac{1}{2}\lambda^2\sigma_\nu^2 + \ln\alpha)(t-1))} \\ \Rightarrow \quad b_t &= \frac{1}{\exp(\lambda\mu + \frac{1}{2}\lambda^2\sigma_\nu^2 + \ln\alpha)^t} x_t^\lambda \end{aligned}$$

Corresponding to different values taken by λ[70], table 2.5 shows the resulting intrinsic bubble processes if fundamentals follow a geometric random walk. It can easily be seen that these intrinsic bubble processes are very similar to those obtained above and tabulated in table 2.4. The remarks mentioned above apply analogously including the multiplicity problem pointed out in the preceding case[71].

Table 2.5: Set of intrinsic bubble processes (geometric random walk)

arbitrary μ, σ_ν^2 arbitrary λ	G1: General intrinsic bubble $b_t = \exp(\lambda\mu + \frac{1}{2}\lambda^2\sigma_\nu^2 + \ln\alpha)^{-t} x_t^\lambda$
arbitrary μ, σ_ν^2 $\lambda = 0$	G2: Deterministic bubble $b_t = \frac{1}{\alpha^t} b,\ b = 1$
arbitrary μ, σ_ν^2 $\lambda = -\frac{2\mu}{\sigma_\nu^2}$	G3: Geometric intrinsic bubble I $b_t = \frac{1}{\alpha^t} x_t^{-\frac{2\mu}{\sigma_\nu^2}}$
$\mu = -\frac{\sigma_\nu^2}{2}$ $\lambda = -\frac{2\mu}{\sigma_\nu^2} \Rightarrow \lambda = 1$	G3a: Restricted geometric intrinsic bubble I $b_t = \frac{1}{\alpha^t} x_t$
arbitrary μ, σ_ν^2 λ s.t. $\lambda\mu + \frac{1}{2}\lambda^2\sigma_\nu^2 + \ln\alpha = 0$	G4: Geometric intrinsic bubble II $b_t = x_t^{\lambda_{1,2}}$

To conclude, by applying the rather general procedure described above, a significant increase in intrinsic bubble processes could be achieved if funda-

[70] Alternatively, the fundamental process (2.49) can be substituted directly into the bubble specifications of table 2.3.

[71] In particular, notice that the restricted geometric intrinsic bubble I when fundamentals follow a geometric random walk, equals the bubble derived for (pure) random walks in equation (2.47).

mentals can be described as a random walk. Some of the resulting bubble processes may form the basis for sucessful investigation of rational bubbles in the data. In chapter 5 we will apply the seemingly most promising ones.

2.4.2.2 Fundamentals as ARMA-processes

Besides these non-stationary random walk processes, ARMA-fundamentals should be examined. However, the procedure for deriving intrinsic bubbles changes significantly in this context. Recall the GSTM process initially introduced, i.e. $M_t = \rho_t M_{t-1} + \zeta_t \epsilon_t$. Until now we have derived a set of bubble processes by restricting ζ_t to be equal to 1. In order to deduce the ARMA-intrinsic bubbles, we now turn to the alternative case by restricting ρ_t to be equal to 1.

CASE 2: $\rho_t = 1$

Unfortunately, it proves difficult to solve for the whole set of ARMA-intrinsic bubbles by using an approach similar to the one demonstrated below. Here, we will apply this general procedure only for the rather simple case of fundamentals following an AR(1)-process. Subsequently, an alternative technique for obtaining intrinsic bubbles will be introduced mainly relying on the already applied method of undetermined coefficients. It turns out that this technique is easier to apply for deriving the whole set of ARMA-instrinsic bubbles. This technique is summarized in proposition 2.13.

Consider the martingale process, i.e. $M_t = M_{t-1} + \zeta_t \epsilon_t$ with $\epsilon_t \sim IID(0, \sigma_\epsilon^2)$[72] and ζ_t representing an arbitrary time dependent random variable. Suppose the fundamental process can be described as an AR(1)-process, i.e. $x_t = \phi x_{t-1} + \epsilon_t$ with $|\phi| < 1$ and $\epsilon_t \sim IID(0, \sigma_\epsilon^2)$. Notice that the error term ϵ_t in both equations is assumed to be equal. Solving both equations for ϵ_t and equating yields

$$
\begin{aligned}
M_t - M_{t-1} &= \zeta_t (x_t - \phi x_{t-1}) \\
&\Rightarrow \begin{cases} M_t = \zeta_t x_t \\ M_{t-1} = \zeta_t \phi x_{t-1} \end{cases}
\end{aligned}
$$

[72]Notice that the white noise property, instead of simple $E[\epsilon_{t+1} \mid I_t] = 0$, is necessary here.

It can easily be seen that the last two equations are satisfied if $\zeta_{t-1} = \phi\zeta_t$ which can be solved for ζ_t resulting in $\zeta_t = \frac{1}{\phi^t}$. The related martingale can then be expressed as

$$M_t = \frac{1}{\phi^t}x_t$$

Recalling the relation between martingales and bubbles in equation (2.41) the corresponding intrinsic bubble is

$$b_t = \frac{1}{(\alpha\phi)^t}x_t$$

establishing the intrinsic bubble specification for AR(1)–fundamentals. Unfortunately, for more general ARMA–processes it becomes more difficult to derive the corresponding intrinsic bubble specification. However, as will be shown later, ζ_t corresponds generally to the weighted sum of the roots of the characteristic equation related to ARMA–processes, i.e.

$$\zeta_t = \textstyle\sum_{i=1}^{p} c_i r_i^t$$
$$r_i \mid \phi_p r^p + \phi_{p-1} r^{p-1} + \ldots + \phi_1 r + 1 = 0$$

where p is the degree of the autoregressive part of the ARMA–process. Thus, the procedure introduced above becomes very uncomfortable for more general ARMA–processes. That is the reason why we present a more facile technique for obtaining martingales, and simultaneously the corresponding intrinsic bubble processes for fundamentals following an ARMA–process.

Hence, suppose the fundamental process can be described as an ARMA–process, i.e.

$$\begin{aligned} \Phi(L)x_t &= \Theta(L)\epsilon_t \\ (1 + \phi_1 L + \ldots + \phi_p L^p)x_t &= (1 + \theta_1 L + \ldots + \theta_q L^q)\epsilon_t \end{aligned}$$

Notice that $\Phi(L)$ is of order p and $\Theta(L)$ of order q. Then the following proposition can be proven.

Proposition 2.13 If the fundamental process can be described as an ARMA(p,q)–process, then the corresponding intrinsic bubble solution has the following form

$$b_t = \frac{1}{\alpha^t}(\Gamma_t(L)x_t + \Delta_t(L)\epsilon_t)$$

$\Gamma_t(L)$ is a polynomial of degree $p-1$ of time dependent coefficients $\gamma_j(t)$

$$\gamma_j(t) = \sum_{i=1}^{p} c_{ji} r_i^t \quad \forall j = 0 \ldots p-1$$

Each coefficient consists of p roots corresponding to the characteristic equation $\phi_p r^p + \phi_{p-1} r^{p-1} + \ldots + \phi_1 r + 1 = 0$ while c_{ij} is determined by initial conditions. $\Delta_t(L)$ is a polynomial of degree $q-1$ consisting of time dependent coefficients being a linear combination of leads of $\gamma_0(t+1)$.

Proof of proposition 2.13 It only must be shown that $\Gamma_t(L)x_t + \Delta_t(L)\epsilon_t$ is a martingale when fundamentals follow $\Phi(L)x_t = \Theta(L)\epsilon_t$. According to the method of undetermined coefficients, a guess concerning a martingale solution has to be formed initially. Thus, our guess is

$$M_t = \Gamma_t(L)x_t + \Delta_t(L)\epsilon_t \tag{2.50}$$

where $\Gamma_t(L)$ is a polynomial of order $p-1$ with time dependent coefficients, i.e. $\Gamma_t(L) = \gamma_0(t) + \gamma_1(t)L + \ldots + \gamma_{p-1}(t)L^{p-1}$ and $\Delta_t(L)$ denotes a polynomial of order $q-1$ with time dependent coefficients, i.e. $\Delta_t(L) = \delta_0(t) + \delta_1(t)L + \ldots + \delta_{q-1}(t)L^{q-1}$.

The solution procedure succeeds as follows: First, leading once and taking expectations of the martingale guess (2.50) yields

$$\begin{aligned} E[M_{t+1} \mid I_t] &= E[(\gamma_0(t+1) + \gamma_1(t+1)L + \ldots \\ &\quad +\gamma_{p-1}(t+1)L^{p-1})x_{t+1} \\ &\quad +(\delta_0(t+1) + \delta_1(t+1)L + \ldots \\ &\quad +\delta_{q-1}(t+1)L^{q-1})\epsilon_{t+1} \mid I_t] \\ &= \gamma_0(t+1)E[x_{t+1} \mid I_t] \\ &\quad +(\gamma_1(t+1)L + \ldots + \gamma_{p-1}(t+1)L^{p-1})x_{t+1} \\ &\quad +(\delta_1(t+1)L + \ldots + \delta_{q-1}(t+1)L^{q-1})\epsilon_{t+1} \end{aligned} \tag{2.51}$$

Second, leading once and taking expectations of the ARMA–fundamentals yields

$$\begin{aligned} E[x_{t+1} \mid I_t] &= -(\phi_1 L + \phi_2 L^2 + \ldots \phi_p L^p)x_{t+1} \\ &\quad +(\theta_1 L + \theta_2 L^2 + \ldots + \theta_q L^q)\epsilon_{t+1} \end{aligned} \tag{2.52}$$

Third, after substituting equation (2.52) into equation (2.51) and taking into account the martingale property $E[M_{t+1} \mid I_t] = M_t$, we obtain

$$
\begin{aligned}
E[M_{t+1} \mid I_t] &= -\gamma_0(t+1)(\phi_1 + \phi_2 L + \ldots \phi_p L^{p-1})x_t \\
&\quad +\gamma_0(t+1)(\theta_1 + \theta_2 L + \ldots + \theta_q L^{q-1})\epsilon_t \\
&\quad +(\gamma_1(t+1) + \ldots + \gamma_{p-1}(t+1)L^{p-2})x_t \\
&\quad +(\delta_1(t+1) + \ldots + \delta_{q-1}(t+1)L^{q-2})\epsilon_t \\
&\stackrel{!}{=} (\gamma_0(t) + \gamma_1(t)L + \ldots + \gamma_{p-1}(t)L^{p-1})x_t \\
&\quad +(\delta_0(t) + \delta_1(t)L + \ldots + \delta_{q-1}(t)L^{q-1})\epsilon_t \\
&= M_t
\end{aligned}
\tag{2.53}
$$

Fourth, two systems of simultaneous equations are obtained from (2.53) by comparing coefficients, one concerning the time dependent coefficient set of $\Gamma_t(L)$ and one concerning the time dependent coefficient set of $\Delta_t(L)$. According to (2.53), the first system of simultaneous equations has the following form

$$
\begin{aligned}
-\gamma_0(t+1)\phi_1 + \gamma_1(t+1) &= \gamma_0(t) \\
-\gamma_0(t+1)\phi_2 + \gamma_2(t+1) &= \gamma_1(t) \\
&\ldots \\
-\gamma_0(t+1)\phi_{p-1} + \gamma_{p-1}(t+1) &= \gamma_{p-2}(t) \\
-\gamma_0(t+1)\phi_p &= \gamma_{p-1}(t)
\end{aligned}
$$

This system of simultaneous equations in matrices can be rewritten as follows

$$
\begin{bmatrix}
-\phi_1 & 1 & 0 & 0 & \cdots & 0 \\
-\phi_2 & 0 & 1 & 0 & \cdots & 0 \\
\vdots & \vdots & \vdots & \vdots & \ddots & \vdots \\
-\phi_{p-1} & 0 & 0 & 0 & \cdots & 1 \\
-\phi_p & 0 & 0 & 0 & \cdots & 0
\end{bmatrix}
\begin{bmatrix}
\gamma_0(t+1) \\ \gamma_1(t+1) \\ \vdots \\ \gamma_{p-2}(t+1) \\ \gamma_{p-1}(t+1)
\end{bmatrix}
=
\begin{bmatrix}
\gamma_0(t) \\ \gamma_1(t) \\ \vdots \\ \gamma_{p-2}(t) \\ \gamma_{p-1}(t)
\end{bmatrix}
$$

or equivalently

$$
\underset{[p \times p]}{\boldsymbol{A}} \quad \underset{[p \times 1]}{\boldsymbol{\gamma(t+1)}} = \underset{[p \times 1]}{\boldsymbol{\gamma(t)}}
\tag{2.54}
$$

We obtain a system of homogenous difference equations which can be solved by using the corresponding characteristic function, i.e.

$$|\boldsymbol{Ar} - \boldsymbol{I}| = 0$$

where **A** is the matrix defined above, **r** is a $[p \times 1]$–vector of roots and **I** denotes the $[p \times p]$–identity matrix. The corresponding characteristic function is a polynomial of degree p and has the following form

$$r_i \mid \phi_p r^p + \phi_{p-1} r^{p-1} + \ldots + \phi_1 r + 1 = 0$$

such that the resulting coefficients can be described as

$$\gamma_j(t) = \sum_{i=1}^{p} c_{ji} r_i^t \quad \forall j = 0 \ldots p-1$$

with c_{ij} to be determined by initial conditions[73].

After once having obtained the solution corresponding to (2.54), the second system of simultaneous equations can be derived from (2.53) again by comparing the coefficients, i.e.

$$\begin{aligned}
\gamma_0(t+1)\theta_1 + \delta_1(t+1) &= \delta_0(t) \\
\gamma_0(t+1)\theta_2 + \delta_2(t+1) &= \delta_1(t) \\
&\ldots \\
\gamma_0(t+1)\theta_{q-1} + \delta_{q-1}(t+1) &= \delta_{q-2}(t) \\
\gamma_0(t+1)\theta_q &= \delta_{q-1}(t)
\end{aligned}$$

which is a recursive system with the following solution

$$\begin{aligned}
\delta_{q-1}(t) &= \theta_q \gamma_0(t+1) \\
\delta_{q-2}(t) &= (\theta_{q-1} + \theta_q L^{-1})\gamma_0(t+1) \\
\ldots & \\
\delta_1(t) &= (\theta_2 + \theta_3 L^{-1} + \ldots + \theta_q L^{-(q-2)})\gamma_0(t+1) \\
\delta_0(t) &= (\theta_1 + \theta_2 L^{-1} + \ldots + \theta_{q-1} L^{-(q-2)} + \theta_q L^{-(q-1)})\gamma_0(t+1)
\end{aligned}$$

characterizing the time dependent coefficients $\delta_j(t)$ as linear combinations of leads of $\gamma_0(t+1)$. ||

[73]Notice that $\gamma_j(t)$ equals the impulse response function of the ARMA(p,q)–process.

Example An example will be useful here. Suppose the fundamental process can be considered as following an ARMA(2,2)–process, i.e.

$$(1+\phi_1 L+\phi_2 L^2)x_t = (1+\theta_1 L+\theta_2 L^2)\epsilon_t$$

The guess concerning the martingale has the following form

$$M_t = (\gamma_0(t)+\gamma_1(t)L)x_t + (\delta_0(t)+\delta_1(t)L^{-1})\epsilon_t$$

In order to determine the time dependent coefficients, we constitute the following two systems of equations: The first system corresponding to $\Gamma_t(L)$ is

$$\begin{bmatrix} -\phi_1 & 1 \\ -\phi_2 & 0 \end{bmatrix} \begin{bmatrix} \gamma_0(t+1) \\ \gamma_1(t+1) \end{bmatrix} = \begin{bmatrix} \gamma_0(t) \\ \gamma_1(t) \end{bmatrix}$$

The characteristic equation (2.55) defines the two corresponding roots r_1 and r_2

(2.55) $$\phi_2 r^2 + \phi_1 r + 1 = 0 \quad \Rightarrow \quad r_1,\ r_2$$

The solution of the first system is, therefore,

$$\begin{aligned} \gamma_0(t) &= c_{11}r_1^t + c_{12}r_2^t \\ \gamma_1(t) &= c_{21}r_1^t + c_{22}r_2^t \end{aligned}$$

where c_{ij} denotes arbitrary coefficients to be determined by initial conditions. After establishing the time dependent coefficients of $\Gamma_t(L)$, the solution of the second (recursive) system is

$$\begin{aligned} \delta_1(t) &= \theta_2 \underbrace{\left(c_{11}r_1^{t+1} + c_{12}r_2^{t+1}\right)}_{=\gamma_0(t+1)} \\ \delta_0(t) &= (\theta_1+\theta_2 L^{-1})\gamma_0(t+1) \\ &= \theta_1(c_{11}r_1^{t+1} + c_{12}r_2^{t+1}) + \theta_2(c_{11}r_1^{t+2} + c_{12}r_2^{t+2}) \end{aligned}$$

Therefore, the resulting martingale for an ARMA(2,2)–fundamental process is

(2.56) $$\begin{aligned} M_t &= (c_{11}r_1^t + c_{12}r_2^t)x_t + (c_{21}r_1^t + c_{22}r_2^t)x_{t-1} \\ &\quad +(\theta_1(c_{11}r_1^{t+1} + c_{12}r_2^{t+1}) + \theta_2(c_{11}r_1^{t+2} + c_{12}r_2^{t+2}))\epsilon_t \\ &\quad +\theta_2(c_{11}r_1^{t+1} + c_{12}r_2^{t+1})\epsilon_{t-1} \end{aligned}$$

which can be rewritten by using $A_t = (c_{11}r_1^t + c_{12}r_2^t)$ and $B_t = (c_{21}r_1^t + c_{22}r_2^t)$ as

$$M_t = A_t x_t + B_t x_{t-1} + (\theta_1 A_{t+1} + \theta_2 A_{t+2})\epsilon_t + \theta_2 A_{t+1}\epsilon_{t-1}$$

By applying relation (2.41) the corresponding intrinsic bubble process can be obtained. The resulting process (2.56) is indeed a martingale corresponding to ARMA(2,2)–fundamentals since for $A_t = (c_{11}r_1^t + c_{12}r_2^t)$ and $B_t = (c_{21}r_1^t + c_{22}r_2^t)$

$$\begin{aligned} E[M_{t+1} \mid I_t] &= A_{t+1}E[x_{t+1} \mid I_t] + B_{t+1}x_t + \theta_2 A_{t+2}\epsilon_t \\ &= A_{t+1}(-\phi_1 x_t - \phi_2 x_{t-1} + \theta_1 \epsilon_t + \theta_2 \epsilon_{t-1}) + B_{t+1}x_t + \theta_2 A_{t+2}\epsilon_t \\ &\stackrel{!}{=} A_t x_t + B_t x_{t-1} + (\theta_1 A_{t+1} + \theta_2 A_{t+2})\epsilon_t + \theta_2 A_{t+1}\epsilon_{t-1} = M_t \end{aligned}$$

Comparing coefficients concerning the AR-part we get the following system

$$\begin{bmatrix} -\phi_1 & 1 \\ -\phi_2 & 0 \end{bmatrix} \begin{bmatrix} A_{t+1} \\ B_{t+1} \end{bmatrix} = \begin{bmatrix} A_t \\ B_t \end{bmatrix}$$

which is the same system of simultaneous equations obtained above for $\gamma_j(t)$ establishing the equivalence. Comparing the coefficients for the MA–part the equivalence is obvious[74].

To conclude, the solution set of intrinsic bubbles related to ARMA–fundamentals is rather complex. The resulting coefficients are highly time dependent and, therefore, difficult to estimate except for the easiest case of an AR(1)–process which can eventually form the basis of testing for the presence of intrinsic bubbles.

2.4.3 Extrinsic bubbles

The idea of extrinsic bubbles is to consider an arbitrary stochastic process $\{z_t, t \in \mathcal{T}\}$ defined outside the model which is supposed to have an impact on the endogenous variable. These variables are usually taken into account through enlarging the information set $I_t = \{\ldots z_t, z_{t-1} \ldots\}$ now additionally containing realizations of past values of z_t. The derivation of

[74]Equivalently, for an AR(1)–processes the resulting characteristic equation yields $r = \frac{1}{\phi_1}$. The result $M_t = \frac{1}{\phi^t}x_t$ already derived above can easily be established by applying proposition 2.13, as well.

extrinsic bubbles goes along the lines presented above. Hence, assuming a special stochastic process $\{z_t, t \in \mathcal{T}\}$, the corresponding results of the preceding section apply. For instance, if z_t follows a random walk, table 2.4 and 2.5 are valid after simple reformulation. If z_t can be described as an ARMA–process, proposition 2.13 provides the ARMA–extrinsic bubble set. The repetition of the process of deriving solutions for extrinsic bubbles is omitted here[75].

When considering extrinsic bubbles, any possibly random variable can influence the movement of the endogenous stochastic process. Although their possibility is easily established by extending the analysis adapted above, extrinsic bubbles have not been examined in literature yet. This may be due to the difficulty in deciding which extrinsic variable should be considered. But also the similarity to a corresponding concept, that of focal point or sunspots, obstruct their empirical examination.

The formal description of sunspots is provided by Cass, Shell (1983) in an exchange economy[76]. They characterize sunspots as economically unreasonable factors which do not influence any variable underlying the exchange model like individuals' utility function, the production technology or individuals' endowment which are, however, thought of having a significant impact on the corresponding rational expectations equilibrium[77].

[75] A special characterization of extrinsic bubbles sometimes encountered in literature should be presented here explicitly serving as an example for a very simple extrinsic bubble. Consider the expectation of an arbitrary random variable z conditional on the information set I_t, i.e. $M_t = E[z \mid I_t]$. Then, M_t forms a martingale since by the law of iterated expectations we get $E[M_{t+1} \mid I_t] = E[E[z \mid I_{t+1}] \mid I_t] = E[z \mid I_t] = M_t$. By substitution of M_t into equation (2.41), the corresponding extrinsic bubble process takes the form $b_t = \frac{1}{\alpha^t} E[z \mid I_t]$. However, notice that this bubble specification can also simply be derived by taking expectations of the deterministic bubble B2. It turns out that this special kind of extrinsic bubble also appears to be contained in the set of bubbles deduced above.

[76] Moreover, sunspots are examined by Azariadis (1981), Spear (1984), Guesnerie (1986) and Azariadis, Guesnerie (1986) among others. For a first overview see Camerer (1989) and Blanchard, Fischer (1989) p 255.

[77] See Azariadis (1981), Spear (1984) and the literature cited in footnote 76 for a derivation of conditions under which sunspots occur. Among these, the incompleteness of markets, i.e. the inability to insure against sunspot events is the most important condition.

In general, sunspots are caused by self–fulfilling beliefs[78] just like rational

Table 2.6: Relationship between bubbles and sunspots

<table>
<tr><th colspan="3">Bubbles</th><th></th></tr>
<tr><td>Markovian bubbles</td><td>Intrinsic bubbles</td><td>Extrinsic bubbles</td><td></td></tr>
<tr><td></td><td></td><td>Explosive sunspots</td><td>Non–explosive sunspots</td></tr>
<tr><td></td><td></td><th colspan="2">Sunspots</th></tr>
</table>

bubbles[79], however, do not necessarily imply explosive behaviour[80]. Hence, the notion of extrinsic bubbles is well captured by the concept of sunspots, however, the latter proves to be broader. The precise relationship between both concepts is clarified in table 2.6 in greater detail. Although sunspots can theoretically be derived from a rational expectations framework, the existence of sunspots like the presence of extrinsic bubbles has not been examined empirically yet, and is left for further research.

2.4.4 Theoretical caveats

Before discussing the economic content of equation (2.4) more precisely, some comments on the reasonability of the bubble concept in general will be necessary. Simultaneously, the aim here is to clarify the framework under which any further analysis of rational bubbles will apply. In particular, besides less binding partial equilibrium arguments for ruling out rational bubbles[81] there exists a general equilibrium argument mainly due to Tirole (1982, 1985), and further extended by Weil (1987, 1989), characterizing the circumstances where rational bubbles may or may not arise. Further theoretical restrictions concerning the existence of rational bubbles pointed out

[78]See Azariadis (1981).

[79]See subsection 3.1.2 for a deeper analysis of rational bubbles induced by self–fulfilling expectations.

[80]See Blanchard, Fischer (1989) p 255.

[81]For instance, a company observing a price bubble related to its stock has the incentive to issue more stocks in order to generate capital gains. However, it is very unlikely that the market will absorb an ever increasing supply of this stock at a higher price than fundamentally justified. So, if supply is arbitrarily elastic, the probability of the existence of rational bubbles at all decreases.

by Diba, Grossman (1988b, 1988c) will be evaluated thereafter. Whenever appropriate, some counterarguments for relaxing these restrictions will be provided[82].

Characterizing the *general equilibrium setting* under which rational bubbles are likely to occur[83], two properties of intertemporal general equilibrium models prove to be important, i.e. the number of individuals and the planning horizon. The easiest framework is that of a finite number of individuals with finite planning horizons. In this case, no bubbles are possible: If a bubble is present, the actual asset price y_t exceeds the fundamental price y_t^f. An individual who buys the asset at $y_t > y_t^f$ will capitalize his capital gain at some finite point of time. Selling the asset at a bubble price, however, will leave other traders with a negative–sum game. Anticipating this, nobody will initially have an incentive to buy the bubble at a bubble price. A similar argument applies if a finite number of individuals with an infinite planning horizon is considered . Tirole (1982, 1990) and Brock (1982) show more rigorously that in this case rational bubbles are ruled out.

Consequently, the crucial assumption refers to the number of individuals in an economy[84]. In his seminal paper, Tirole (1985)[85] relying on a perfect foresight overlapping–generation setting, in which the infinite chain of individuals implies the infinite number of market agents, demonstrates that deterministic rational bubbles are possible if the growth rate of the economy n is greater or equal to the growth rate of the bubble r, indicating

[82]There are other prerequisits for deriving rational bubbles. For instance, the asset considered must live forever. If not, bubbles cannot exist. Since at the end of an asset's lifetime the bubble bursts with certainty, individuals knowing this would never buy the asset above its fundamental value in the period before. The same argument applies two periods before since knowing the bubble will burst with certainty in $T-1$ prevents one from buying the asset above its fundamental value in $T-2$. By extension, a bubble can never exist in assets with finite lifetimes (like bonds). Further, free disposal of assets rules out rational bubbles. For further arguments see Blanchard, Fischer (1989) p 222.

[83]A good overview provides Diba (1990).

[84]There is often a confusion between these two characterizations of general equilibrium models. Blanchard, Fischer (1989), for instance, do not distinguish clearly between these two concepts stating that the finite–horizon argument is the decisive one.

[85]See also Tirole (1990).

a dynamically inefficient economy. This condition can be rationalized by considering that if this condition is violated, the bubble may become larger than the economy itself and nobody will be able to buy the asset containing that bubble[86]. Weil (1987) examines the possibility of bursting bubbles

Table 2.7: General equilibrium arguments

	Infinite planning horizon	Finite planning horizon
Finite number of individuals	No bubbles possible Tirol (1982), Brock (1982)	No bubbles possible
Infinite number of individuals	Bubbles possible Weil (1989)	Bubbles possible Tirole (1985), Weil(1987)

and yields only a slightly different result. Recently in his paper (1989), he extended his overlapping generations model by additionally considering a bequest motive, which links subsequent generations together and results in an infinite planning horizon of the individuals. Weil (1989) succeeds in deriving rational bubbles in this setting, as well[87]. To conclude, the theoretical general equilibrium framework necessary for the occurrence of rational bubbles is summarized in table 2.7.

Another restriction on the appearance of rational bubbles is the *non-negativity property*. Within a monetary framework Brock (1974, 1975), Gray (1984) and Obstfeld, Rogoff (1983, 1986) exclude deflationary price level bubbles by applying appropriate transversality conditions[88]. The economic relevance of this property simply lies in the fact that if negative bubbles are allowed, they are expected to grow to minus infinity, violating the positiveness of prices.

Based on this non-negativity property, Diba, Grossman (1988b, 1988c) present two further properties of rational bubbles: First, if rational bubbles are present, they must have been present from the first day of trading the asset, and second, a bubble that once has burst cannot restart.

[86]However, recently Diba (1990) constructed a general equilibrium framework where rational bubbles are possible even if the interest rate exceeds the population growth. This result is mainly due to a special tax structure imposed by Diba (1990).

[87]See O'Connell, Zeldes (1988) for a similar outcome.

[88]See section 2.4.4 for a further analysis.

The *first property* can be shown by applying proposition 2.12 stating that bubbles follow $E[b_{t+1} \mid I_t] = \frac{1}{\alpha}b_t$ for $0 < \alpha < 1$[89]. Based on the non-negativity property of rational bubbles, i.e. $b_t \geq 0$, we get by introducing a rational prediction error ϵ_{t+1}

$$\begin{aligned} b_{t+1} &= \frac{1}{\alpha}b_t + \epsilon_{t+1} \geq 0 \\ \Rightarrow \qquad \epsilon_{t+1} &\geq -\frac{1}{\alpha}b_t \end{aligned}$$

Hence, with non-negative rational bubbles the rational prediction error ϵ_{t+1} must exceed $-\frac{1}{\alpha}b_t$ in every period. If $b_t = 0$ for an arbitrary t, the rational prediction error must be greater than zero in order to prevent negative rational bubbles. However, the rational prediction error is equally likely to be negative or positive due to its mean zero property. Consequently, the rational prediction error takes the value zero with probability one in order to guarantee the non-negativity of asset prices. This argument applies for all t and also by extension for $t = 0$. In order to establish the possibility of rational bubbles at all, a rational bubble must have been present from the first day of the existence of the corresponding asset[90].

The *second argument* goes similarily. Extending the above argument suppose a bursting bubble

$$b_t = \begin{cases} \frac{1}{q\alpha}b_{t-1} + u_t & \text{with probability } \pi \\ u_t & \text{with probability } 1 - \pi \end{cases}$$

with $E[u_t \mid I_{t-1}] = 0$. After crashing the bubble process equals u_t which is zero on average. Applying the same argument as above in the case of bursting bubbles, u_t must take the value zero with probability one in order to avoid negative bubbles, i.e. negative asset prices. Hence, Diba, Grossman (1988b, 1988c) conclude that rational bubbles must have been started at the first day of the existence of the corresponding asset and cannot restart once crashed.

The arguments described above mainly rely on the assumption of non-negative bubbles. However, the true restriction concerns the endogenous

[89]This restriction is imposed in economic applications and will be justified in the next chapter.

[90]Notice that this implies that a stock, for instance, is expected to be overevaluated by the individuals from the first day of trading. Otherwise no bubble would ever exist.

process (asset prices, for instance) to be non-negative. Negative bubbles may then be neutralized by fundamentals growing at a sufficient rate. But it can be shown that if this is the case, essential parameter restrictions are violated. Recall the two components of the general solution of equation (2.4) obtained earlier, i.e the fundamental f_t and the bubble solution b_t. According to proposition 2.12, the bubble grows at rate $-\ln\alpha$. Assuming $\alpha = \exp(-r)$ with r as the real interest rate, which is common in many economic applications, r can be viewed as the growth rate of the bubble. Preventing the endogenous process y_t from taking negative values requires that the fundamental component f_t should grow at a rate greater than r if $b_0 < 0$. Suppose the exogenous process x_t follows a geometric random walk with growth rate μ. Then, if $\mu > r$, negative bubbles may be possible since the growth of the fundamental solution compensates the decrease in the bubble component[91]. However, notice that this parameter constellation is excluded since by proposition 2.10 the parameters must satisfy $\mu + \frac{\sigma^2}{2} < r$ for the convergence of the present value relation, i.e. the fundamental solution. This implies $\mu < r$ in every case confirming the impossibility of negative bubbles.

While the restriction of non-negative rational bubbles cannot be relaxed, it can easily be shown that periodically bursting or shrinking bubble processes may exist without violating the non-negativity property of prices. Diba, Grossman (1988b, 1988c) themselves have introduced such a bubble process[92]. For instance, the whole set of intrinsic bubbles, while periodically decreasing, never vanishes completely as long as the fundamental process takes a value different from zero. Moreover, other processes like B1 containing an exponential term also avoid becoming negative, although being able to exhibit a bursting-like behaviour. Notice that in this setting 'bursting' does not mean that the value of the bubble becomes entirely zero[93]. Here, the bubble becomes observationally undistinguishable from completely bursting bubbles by showing a bursting-like behaviour.

---

[91]The analysis of Köddermann (1989) relies mainly on this arguments, however, it neglects the parameter restrictions related with the fundamental solution.

[92]Evans (1991) presents another example.

[93]In the case of an entire crash of the bubble the result of Diba, Grossman (1988b, 1988c) naturally remains valid.

Finally, in order to investigate the possibility of rational bubbles and the constraints connected with them, Allan, Gorton (1991, 1993) recently chose a different approach. They apply a principal–agent setting in which investors have asymmetric information concerning the performance of portfolio managers. In this framework it can be shown that bubbles can crash and restart even though the number of individuals is finite. The reason for this lies mainly in the fact that traders cannot figure out whether there are other traders left in the market when considering buying an asset since they do not know which traders bought the asset before. Hence, this paper has cast doubt on the relevance of at least the general equilibrium argument, and the restarting argument, which appear to depend crucially on the type of model considered.

2.4.5 Summary

In this section we have distinguished three different classes of bubble processes. First, the Markovian bubble processes have been examined, in particular a deterministic, a random walk and a bursting bubble. Second, within the class of intrinsic bubbles characterized as depending merely on fundamentals, bubble processes related to different stochastic representations of the fundamental processes have been obtained. Non–stationary random walk and stationary ARMA–specifications of fundamentals have been investigated. Finally, the behaviour of the third class of rational bubbles, i.e. extrinsic bubbles, is shown to show very similar behaviour to that of intrinsic bubbles. The different bubble processes discussed so far are summarized in table 2.8. It becomes obvious that only a part of the special bubble processes, displayed in table 2.8, has been examined in literature so far. Especially, the set of intrinsic bubbles could have been extended significantly. This gives rise to the hope that the possibility of empirically confirming the existence of rational bubbles will notably increased.

Moreover, a general procedure for deriving bubble processes has been introduced by providing a common origin based on a so–called generalized stochastic martingale (GSTM) process. Consequently, the collection of explicit bubble processes shown in table 2.8 represents only a small subset of bubble processes possibly derivable from a GSTM process. Based on this general procedure, further research may be undertaken in order to increase the set of more sophisticated bubbles.

Finally, some restrictions and arguments for theoretically ruling out bubbles have been provided and discussed. However, except for the non-negativity of rational bubbles, the relevance, at least of some of them, depends highly on the models considered.

Table 2.8: Entire set of bubble processes

	CASE 1: $\zeta_t = 1$	**CASE 2:** $\rho_t = 1$
Markovian bubbles	Table 2.3: **Deterministic bubble**	
	B2: $b_t = \frac{1}{\alpha^t} b, \;\; b > 0$	
	Table 2.3: **Near random walk bubble**	
	B6: $b_t = \frac{1}{\alpha} b_{t-1} + u_t$	
	Bursting bubble	
	$b_t = \begin{cases} \frac{\rho_t}{\pi\alpha} b_{t-1} + u_t & \text{with prob. } \pi \\ u_t & \text{with prob. } 1-\pi \end{cases}$	
Intrinsic bubbles	Eq. (2.47) **(Pure) random walk**	**ARMA-fundamentals**
	RW: $b_t = \frac{1}{\alpha^t} x_t$	Prop. 2.13
	Table 2.4: **Random walk with drift**	**AR(1)-fundamentals**
	I1: $b_t = \exp(\lambda\mu + \frac{1}{2}\lambda^2\sigma_\nu^2 + \ln\alpha)^{-t} \exp(x_t)^\lambda$	$b_t = \frac{1}{(\alpha\phi)^t} x_t$
	I3: $b_t = \frac{1}{\alpha^t} \exp(x_t)^{-\frac{2\mu}{\sigma_\nu^2}}$	
	I3a: $b_t = \frac{1}{\alpha^t} \exp(x_t)$	
	I4: $b_t = \exp(x_t)^{\lambda_{1,2}}$	
	Table 2.5: **Geometric random walk**	
	G1: $b_t = \exp(\lambda\mu + \frac{1}{2}\lambda^2\sigma_\nu^2 + \ln\alpha)^{-t} x_t^\lambda$	
	G3: $b_t = \frac{1}{\alpha^t} x_t^{-\frac{2\mu}{\sigma_\nu^2}}$	
	G3a: $b_t = \frac{1}{\alpha^t} x_t$	
	G4: $b_t = x_t^{\lambda_{1,2}}$	
Extrinsic bubbles	**(Pure) random walk**	**ARMA process**
	analogous to equation (2.47)	analogous to Prop. 2.13
	Random walk with drift	**AR(1) process**
	analogous to Table 2.4	see above
	Geometric random walk	
	analogous to Table 2.5	

Chapter 3

On the Economic Relevance of Rational Bubbles

In the preceding chapter we provided solution procedures applicable to the simple stochastic difference equation

$$y_t = \alpha E[y_{t+1} \mid I_t] + x_t \tag{3.1}$$

The purpose of this chapter is to enrich equation (3.1) with economic content. Although equation (3.1) seems to be a rather simple one, it is often encountered in economic theory[1]. In the following we will examine the relevance of equation (3.1) for explaining the behaviour of asset prices, exchange rates and the price level in periods of hyperinflation. The discussion is mainly concentrated on capital markets since the econometric application in chapter 4 relates to the German Stock Market. The latter two applications of rational bubbles will be considered shortly. However, many aspects of the exposition of rational bubbles on capital markets, primary the 'bubble cycle', can easily be transferred to other interpretations of equation (3.1).

[1]A first exposition on economic applications of equation (3.1) is provided by Heri (1986).

3.1 Capital markets

The aim of this section is to examine the relevance of equation (3.1) in the context of capital markets. A special emphasis lies on the revelation of the underlying assumption structure necessary to derive equation (3.1)[2]. These preconditions will critically be discussed at the end of this section. Moreover, equation (3.1) will be derived from an intertemporal optimization framework providing a microfoundation for further theoretical and empirical examination of rational bubbles on capital markets. Then, both components of the general solution already presented in the preceding chapter will be discussed independently. The fundamental solution, i.e. the present value relation, is often associated with the notion of efficient markets. The relationship between present value models and the efficient market hypothesis will be clarified by discovering the underlying restrictions needed in order to establish the equivalence. The bubble solution will then be interpreted as an additional component capable of describing some stylized facts on capital markets which cannot be explained by simple present value models. Finally, a so–called 'bubble cycle' which tries to illustrate the evolution of rational bubbles in greater detail will be developed.

In order to establish equation (3.1) in the context of capital markets, consider the following definitorial equation for the ex–post, one–period return r_t of a risky asset comprising both a return component, due to a fundamental cash flow related to that asset, and a capital gain, or loss, due to the variability of asset prices. Assuming the absence of transaction costs[3], we get

$$r_t = \frac{p_{t+1} - p_t}{p_t} + \frac{d_t}{p_t} \tag{3.2}$$

where p_t denotes the price of the corresponding asset and d_t the underlying fundamental cash flow. This relationship is always true in an ex–post analysis and, therefore, tautological. Through imposing certain assumptions, equation (3.2) can be extended to a simple financial market model

[2]See LeRoy (1982) among others for a first presentation of the structure of the underlying assumptions of equation (3.1) in various economic applications.

[3]See Grossman, Stiglitz (1980) for a discussion of the relevance of this presumption.

which tries to explain the behaviour of asset prices ex–ante. First, in an uncertain environment for defining the ex–ante rate of return, one requires an assumption concerning the way expectations are formed by individuals.

Assumption 1 Individuals have rational expectations.

Considering rational market participants, the expected one–period return on a risky asset becomes, after taking conditional expectation,

$$E[r_t \mid I_t] = \frac{E[p_{t+1} \mid I_t] - p_t}{p_t} + \frac{d_t}{p_t} \tag{3.3}$$

Notice that d_t is assumed to be known at t, i.e. d_t is already contained in I_t[4]. The notion of rational expectations, as already pointed out above, is often associated with the assumption of a homogenous spread of information including an identical economic model used by all individuals. This becomes relevant when discussing alternatives to the rational expectations model proposed here[5].

In equation (3.3) only expected return components are taken into account. Self–restricting to the first moment of return distributions implies the assumption that risk considerations, usually measured in terms of the second moment, do not play a major role in determining asset prices. Hence, it becomes plain that one of the decisive assumptions related to equation (3.1) concerns the risk perception of individuals involved[6].

Assumption 2 Individuals are risk–neutral.

As it will be shown and discussed more extensively below, the risk–neutrality assumption can also be obtained in an intertemporal optimization setting as a prerequisite for deriving equation (3.1). Risk–neutrality means that

[4]The analysis would not change significantly if d_t is assumed to be unknown at t.

[5]The existence of fundamentalists and chartists may describe a possible deviation from the assumption of the use of identical models.

[6]A more extended version of equation (3.1) presented below when discussing the alternatives additionally implies a constant risk premium. The assumption then must be restated in terms of the constancy of the risk premium in contrast to a risk premium varying with time.

individuals are indifferent between receiving any lottery and the expectation of that lottery. Here the lottery consists of the choice between risky and riskless assets. A direct consequence of this assumption is that both riskless and risky assets will yield the same expected return since if only obtainable returns are relevant for investment decisions, any return differences between assets will be arbitraged.

The last decisive assumption for deriving equation (3.1) is related to the expected rate of return. For stationary rates of return the rationally expected value can be identified with the sample mean $\bar{r}$ if the sample size is large enough for deriving a consistent estimate. Accepting this, we arrive at the following condition.

Assumption 3 The rationally expected rate of return $E[r_t \mid I_t]$ of a risky asset is constant over time, i.e. $E[r_t \mid I_t] = \bar{r} = r$ for $\forall t$.

Taking into account the fact that under risk–neutrality the rate of return on a risky asset equals the return on a riskless asset, the one–period expected rate of return r can simply be identified with the real interest rate.

Combining the assumptions presented above and rearranging the obtained relation adequately, we arrive at the following difference equation

$$p_t = \frac{1}{1+r} E[p_{t+1} \mid I_t] + \frac{d_t}{1+r} \tag{3.4}$$

The equivalence to equation (3.1) is easily established by taking $\alpha = \frac{1}{1+r}$ and $x_t = \frac{d_t}{1+r}$. Notice that in the normal case of positive real (average) interest rates, α becomes smaller than one which is important for deriving the fundamental solution component. Equation (3.4) also represents a no–arbitrage–condition.

The underlying assumption structure becomes more apparent when deriving equation (3.4) from an intertemporal optimization framework[7]. Assume that at time t a representative individual maximizes his expected

[7] Among others see Grossman, Shiller (1981), Flood, Hodrick, Kaplan (1986), Diba, Grossman (1988a), or Engsted (1993a). See Grossman, Stiglitz for a continuous time setting.

utility function depending additive separably on current and rationally expected future consumption c over an infinite time horizon[8]

$$\max_{c_{t+i}} \sum_{i=0}^{\infty} \vartheta^i E[u(c_{t+i}) \mid I_t]$$

where the information set I_t contains all past and current variables of the model and $\vartheta > 0$ represents the time preference of the individual. Optimizing utility through suitable choice of the consumption path, the individual takes into account an investment opportunity on the capital market. The asset considered yields a dividend of d_t at the end of period t increasing the consumption possibility in period $t+1$. Whereas the dividend paid in period t is known, all future dividend payments are uncertain. The individual has the opportunity to buy the amount h_t of the risky asset at the beginning of period t, at known price p_t, and may sell it at unknown price p_{t+1} one period later. Given an exogenous stream of income w_t the individual maxizimes his utility under the following budget constraint for $\forall\, t$

$$c_{t+1} = w_{t+1} + (p_{t+1} + d_t)h_t - p_{t+1}h_{t+1} \tag{3.5}$$

The individual faces two sources of income, i.e. labour income w_t and capital income, due to the investment on capital markets which includes a fundamental component $d_t h_t$ and a possible capital gain. Income can be spent for consumption and for further investment in the risky asset. Substitution of equation (3.5) adequately defined for two successive points of time into the utility function and differentiation with respect to h_t, determines the optimal demand for stocks. The following first order condition (Euler–equation) for all t arises

$$u'(c_t)p_t = \vartheta E[u'(c_{t+1})](E[p_{t+1} \mid I_t] + d_t)$$

which can be transformed into

$$p_t = \vartheta \frac{E[u'(c_{t+1})]}{u'(c_t)}(E[p_{t+1} \mid I_t] + d_t) \tag{3.6}$$

In order to establish the equivalence between equation (3.6) and (3.4), the following condition must hold $\vartheta \frac{E[u'(c_{t+1})]}{u'(c_t)} = \frac{1}{1+r}$. However, if the rate

[8] The assumption of an infinite time horizon is not necessary for deriving equation (3.4) as was pointed out at the end of the last chapter.

of time preference ϑ is assumed to be equal to the discount factor, the condition simplifies to

$$\frac{E[u'(c_{t+1})]}{u'(c_t)} = 1 \tag{3.7}$$

i.e. the ratio of marginal utility from expected future and current consumption must equal one which implies a constant expected rate of substitution. A constant expected rate of substitution, however, at all points of time, and independent of the ratio of expected consumption c_{t+1} and c_t, requires linear indifference curves, i.e. a utility function which depends linearly on consumption at each point of time. How can this condition be met?

- First, by assuming a special characterization of the utility function implying a constant relative risk aversion (CRRA) which has the following form

$$\begin{aligned} u(c_t) &= \frac{c_t^{1-\delta}}{1-\delta} \quad \text{for } \delta \neq 1 \\ u(c_t) &= \ln c_t \quad \text{for } \delta = 1 \end{aligned}$$

  where $\delta \geq 0$ denotes the constant relative risk aversion parameter interpretable as a measure of disutility of consumption fluctuations[9]. The expectational expression can be written as

$$\frac{E[u'(c_{t+1})]}{u'(c_t)} = \left(\frac{c_{t+1}}{c_t}\right)^{\delta}$$

  It can easily be seen that the required restriction (3.7) is satisfied for $\delta = 0$ describing the case of risk-neutrality. Hence, the assumption of risk-neutrality for deriving the stochastic difference equation (3.1) is also obtained in an intertemporal optimization framework with CRRA-utility function when imposing a restriction on the risk attitude parameter. Notice that for $\delta = 0$ the CRRA-utility function becomes linear, a result which has already been theoretically established above.

- Second, another possibility for fulfilling the above restriction is to assume only slight random changes in consumption between two successive periods which may capture the idea of a sort of ratchet-effect

[9] A CRRA-utility function is characterized by $-\frac{u''c}{u'} = \delta = \text{const.}$

in consumption. For instance, suppose consumption simply follows a martingale, i.e.

$$E[c_{t+1} \mid I_t] = c_t$$

then again the above mentioned restriction is satisfied[10]. In this case, equation (3.4) can be established even if individuals are risk averse with constant risk parameter δ.

Under one of the conditions derived above we get for all t

$$p_t = \vartheta E[p_{t+1} \mid I_t] + \vartheta d_t \tag{3.8}$$

Hence, the previously mentioned arbitrage equation (3.4) can easily be obtained from an optimization behavior of a representative consumer. Notice that the equivalence of equation (3.4) and (3.8) prevails only for $\vartheta = \frac{1}{1+r}$ which suggests the correspondence of individual's time preference and the discount factor as already mentioned.

Based on the solution principles derived in the previous chapter, we apply the forward solution as fundamental solution component since α turns out to be smaller than one. Hence, for d_t following a non-prespecified stochastic process, the general solution, including the bubble component, is according to section 2.3.1

$$p_t = \sum_{i=0}^{\infty} \frac{1}{(1+r)^i} E[d_{t+i} \mid I_t] + (1+r)M_t$$

where M_t denotes an arbitrary martingale. Taking a closer look at the fundamental part of the solution, it turns out that asset prices should fundamentally equal the sum of an infinite rationally expected dividend stream discounted back at rate r. The bubble solution component, however, containing a martingale should grow at the rate of interest r in every period.

While the bubble solution was neglected for a long time in economic research, the fundamental component enjoyed widespread attraction. Above

[10] A special case is obtained if consumption can be described by a random walk. The random walk property of consumption is formally derived by Hall (1978).

all, the present value relation derived above often forms the basis for examining the efficiency of markets, especially of capital markets[11]. Besides this line of empirical work on efficient markets as related to present value models, a body of research was associated with other descriptions of efficiency, based mainly on Fama's (1970) proposals for testing of efficiency by labelling prices as following a martingale or random walk, or characterizing the corresponding returns as a fair game. In the following we will briefly show how both empirical approaches are connected. To begin, we shall define the concept of efficient markets; by no means an easy task, and, thereafter, discuss the bubble solution more extensively.

3.1.1 Efficient capital markets

The notion of efficient markets is due to Fama (1970)[12]. In his paper he defines markets as being efficient if market prices fully reflect all information relevant for their evaluation. No additional information can be exploited in order to realize extraordinary gains through trading. Thus, the concept of efficient markets describes a situation where a market is *informationally* efficient. Assuming a special information set I_t, from the definition of efficient markets, it follows immediately that the acquisition of information contained in I_t is needless. I_t provides no extra information which might serve as a basis for profitable trading rules since all information is already incorporated in prices. Moreover, on efficient capital markets expected changes in future earnings of firms are instantaneously reflected in corresponding variations in asset prices by arbitrage transactions. Contrarily, a capital market is not fully efficient if profitable trading rules which are systematically exploitable exist. Notice that trading rules may be profitable by chance although capital markets are efficient.

When defining efficient markets, the underlying information structure becomes important. Besides the assumption of growing information sets

[11]Rational bubbles have usually been ruled out by implicitly assuming an adequate transversality condition.

[12]For a more recent overview of efficient market literature see Fama (1991). There, he summarizes the different lines of research and the current state of the debate over whether capital markets can be viewed as efficient or not.

usually implied by rational expectations, three different characterizations concerning the information set I_t may be distinguished[13]:

1. A market is said to be *weak–form efficient*[14] if the information set I_t contains all current and past realizations of prices (and realizations of cash flows related to that asset) and no trading rule based on I_t can expect to yield systematic gains.

2. Further, a market is said to be *semi–strong efficient*[15] if predictions based on the information set I_t, which in addition to weak–form efficiency includes all publicly available information, generate no excess returns. Public information is comprised of, for instance, annoucements of higher investment of a company, macroeconomic performance measures and general economic and political indicators which respectively are supposed to be relevant for the market.

3. Finally, a market is said to be *strong–form efficient*[16] if additionally private or better insider information is already incorporated into the market price and cannot be exploited by market agents.

Mostly the weak–form efficiency has formed the basis for empirical in-

[13]See Fama (1970). Notice that rational expectations are additionally connected to the assumption of information homogenously spread out among individuals. In his seminal paper on efficient markets Fama (1970) assumed homogenous expectations among individuals. In a later paper Fama (1976) generalizes this implication of rational expectations in the sense that single individuals may possess different information, however, that the aggregated information set representing the market knowledge should contain all relevant information for evaluating the price of assets.

[14]In Fama (1991) this category of efficiency is captured by the concept of *unpredictability of returns*. From recent studies no unique answer can be drawn whether capital markets are efficient or not. However, it seems that returns are likely to be predictable in the short run contradicting the paradigm of market efficiency. See Fama (1991) for a further discussion.

[15]In Fama (1991) this type of efficiency is investigated by so–called *event studies*. Interestingly, most recent empirical studies confirm this form of efficiency.

[16]In Fama (1991) *tests for private information* capture the notion of strong–form efficiency. No clear outcome concerning this kind of efficiency can be drawn from the studies carried out so far. See Fama (1991) for a more detailed description.

vestigations on capital, foreign exchange rate and other asset markets[17]. However, the question arises how can the notion of 'prices fully reflect information' be made operational[18]. Fama (1970) himself proposes to validate the hypothesis of efficient capital markets by showing the martingale or random walk property of asset prices or, alternatively, by describing the related rate of return as a fair game by using correlation tests, runs tests and various filter rule techniques among other test procedures. On the other side, an alternative approach of making the notion of efficient market operational exists which associates efficiency with the validity of the present value model[19]. Obviously two seemingly opposite approaches to formally defining efficient markets exist; one based on martingales and one related to the present value model both relying on the assumption of rational expectations. Here, we will first introduce the concepts of martingales, random walk and fair games and discuss their relevance for efficient capital markets. Then, we shall show how these models are connected to the present value relation.

One of the earliest models[20] related to capital markets is the *random walk* model. The idea is that changes in asset prices are not predictable since they are viewed as fluctuating only randomly. This implies that the best forecast of tomorrow's asset price p_{t+1} is today's price p_t plus an error term ϵ_{t+1} identically and independently distributed with zero mean and constant variance, i.e.

$$p_{t+1} = p_t + \epsilon_{t+1}$$

[17]Notice that in the last chapter we used the weak-form characterization of I_t when deriving the fundamental solution. Considering additionally the bubble solution, the information set must possibly be enlarged. For instance, examining the impact of extrinsic factors publicly available requires an information set satisfying the semi-strong definition of I_t rather than the weak-form.

[18]For a discussion of the empirical counterpart of efficient market hypothesis see LeRoy (1982).

[19]See Shiller (1979, 1981b) and LeRoy, Porter (1981) among others. Section 4.1 discusses this line of research more extensively.

[20]The first paper addressed to the random walk property of asset prices was the dissertation of Bachelier (1900) in which he empirically shows that stock price changes fluctuate randomly. His work was buried for a long time and first attracted public attention when published in Cootner's (1964) exposition on stock markets. Another seminal work related to the random walk property of stock prices is Working (1934).

In Fama's (1970) view this description of the efficient market hypothesis captures well the idea that no systematically profitable trading rules exist. However, the random walk property of asset prices requires serially uncorrelated error terms which appear to be an extreme assertion, at least if capital markets are considered. A more convincing concept is the notion of *martingales*[21] which do not require any assumption concerning the error term beyond the uncorrelatedness of disturbances from expected future prices. On the other side, the concept of martingales, like random walks, is also able to show the property of unpredictability in future price changes. The martingale hypothesis simply states that

$$E[p_{t+1} \mid I_t] = p_t$$

i.e. the expected price tomorrow equals the price today which already contains all prior information. Again, no profitable trading can be systematically based on this description of the behaviour of asset prices. An extension of the concept of martingales provides the notion of submartingales describing the idea of increasing price paths sometimes ascribed to actual data. A time series follows a submartingale if

$$E[p_{t+1} \mid I_t] \geq p_t$$

i.e. the expected price tomorrow is greater than or equal to the price today.

While both the random walk, and the more general martingale approach, try to explain the movement of asset prices themselves, the concept of *fair games* is related to returns on assets. The fair game property can easily be established from the martingale approach by taking logs of asset prices and then differencing once[22] which provides an approximation of the corresponding rate of return. In the case of asset prices describable by a (pure) martingale, we get the following fair game definition

$$E[r_t \mid I_t] = E[\ln p_{t+1} \mid I_t] - \ln p_t = 0$$

---

[21]Notice that here the endogenous process itself follows a martingale. This is to be distinguished from the bubble solution containing a martingale process.

[22]A fair game in the context of capital markets is, also, often defined without taking logs. Then fair games coincide with the concept of *martingale differences*.

i.e. if prices follow a (pure) martingale, the expected rate of return of the corresponding fair game is zero[23]. Assuming prices grow at a certain rate which implies the submartingale–property, the equivalent definition of a fair game can be obtained by taking logs and differencing once

$$E[r_t \mid I_t] = E[\ln p_{t+1} \mid I_t] - \ln p_t \geq 0$$

Here a fair game states that no excess return above a certain predetermined level of return can be expected[24].

In order to characterize the conditions under which the efficient market approach, as related to concepts of martingales, random walks or fair games, and the approach based on the present value relation, coincide, we must show only how the stochastic difference equation (3.4) can be identified with the (sub–)martingale model, unlike having to establish the equivalence of present value models and martingales directly. From equation (3.4) the present value model can be derived by imposing an appropriately defined transversality condition as already shown. The relationship between the different concepts of how to make the notion of market efficiency operational, is displayed in figure 3.1.

Figure 3.1: Relationship between martingales and present value models

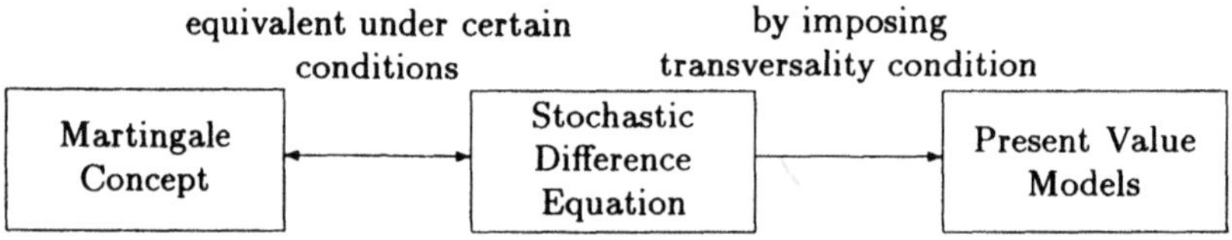

It will turn out that the equivalence of both approaches is only true under restrictive conditions. We concentrate on the martingale model; the relation between martingales and random walk and fair game concepts has already been shown above.

[23] A typical fair game is described by a lottery with two equally likely outcomes. In the first event the gambler wins a certain amount, otherwise he looses the same amount. The expected wealth after playing the game is the same as before.

[24] This level is easily derived below, under the assumption of risk–neutrality, as the dividend–price ratio. Taking into account risk–averse agents, the rate of return expected by the capital asset pricing model, containing a constant risk component related to the market portfolio, may represent this predetermined level, as well.

1. The simplest way to derive the equivalence between the martingale concept for asset prices and the stochastic difference equation (3.4) is to assume that dividend payments are zero. Notice that in this case the intrinsic value of the asset is zero[25]. Hence equation (3.4) becomes

$$p_t = \frac{1}{1+r} E[p_{t+1} \mid I_t]$$

For daily changing asset prices, i.e. for small time steps, the rate of return r is near zero, directly establishing the martingale property of asset prices, i.e. $p_t = E[p_{t+1} \mid I_t]$. Thus, the martingale property of asset prices is derived under the assumption of zero dividend payments and a rate of return being approximately zero[26]. For rates of return greater than zero merely the submartingale property of asset prices $p_t \leq E[p_{t+1} \mid I_t]$ can be established implying that asset prices grow at least at the rate of the riskless asset[27].

2. More generally, admitting non-zero dividend payments, Samuelson (1973)[28] shows that the martingale property of asset prices and the stochastic difference equation are only equivalent for a special rate of return r. Based on equation (3.4), i.e. $p_t = \frac{1}{1+r} E[p_{t+1} \mid I_t] + \frac{d_t}{1+r}$ and exploiting the martingale property $p_t = E[p_{t+1} \mid I_t]$ an expression can be derived which determines the required rate of return in order to reach the equivalence of both approaches. After simple rearrangements we get

$$\begin{aligned} p_t &= \frac{1}{1+r} p_t + \frac{d_t}{1+r} \\ \Leftrightarrow \quad r &= \frac{d_t}{p_t} \end{aligned} \tag{3.9}$$

Thus, asset prices follow a martingale and can equivalently be described by equation (3.4) if the rate of return r equals the dividend–price ratio. Notice that this is the standard definition of returns

---

[25]This can be reasonable for intrinsically useless assets as money, for instance.

[26]See, for instance, Hansen (1993) p 359.

[27]Moreover, in his (1965) paper Samuelson proves that the price of futures in its simplest form follows a martingale and a submartingale when a discount factor is included.

[28]For a good presentation of Samuelson's ideas in the context of financial markets see, for instance, Brock, Malliaris (1982) p 25 and LeRoy (1989).

relating the revenue connected with an investment (here d_t) to the invested amount of money per unit (here p_t)[29].

Further, equation (3.9) may simultaneously serve as a reference point for determining excess returns when r, defined in equation (3.9), can be viewed as a 'normal' return. Then, asset prices follow a sub(super-)martingale, i.e. $p_t \leq (\geq) E[p_{t+1} \mid I_t]$ if the rate of return is higher (lower) than the dividend–price ratio $r \geq (\leq) \frac{d_t}{p_t}$.

3. For arbitrary rates of return the correspondence of the martingale property and the stochastic difference equation (3.4) can only be established by considering the value of a mutual fund v_t discounted back to date zero which cumulates the distributed dividend payments[30]. Asset prices themselves cannot be described as following a martingale. Thus, define the value of a mutual fund v_t as

$$v_t = \frac{1}{(1+r)^t} p_t h_t \tag{3.10}$$

where h_t stands for the number of shares of the mutual fund at time t. Cumulation of dividend payments can be described as follows

$$p_{t+1} h_{t+1} = (p_{t+1} + d_t) h_t \tag{3.11}$$

The martingale property of v_t can easily be shown. Taking expectations of equation (3.10), using equation (3.4) and (3.11) and simple algebra yields

$$\begin{aligned} E[v_{t+1} \mid I_t] &= E\left[\frac{1}{(1+r)^{t+1}} p_{t+1} h_{t+1} \mid I_t\right] \\ &= \frac{1}{(1+r)^{t+1}} (E[p_{t+1} \mid I_t] + d_t) h_t \\ &= \frac{1}{(1+r)^{t+1}} (1+r) p_t h_t = v_t \end{aligned}$$

[29]Since r is assumed to be constant over time, the expected dividend–price ratio should be constant, as well. Further, under the condition that prices follow a martingale, dividends also must follow a martingale implying that individuals expect dividends to be kept at the pre–period level. This coincides with the observation that managers tend to smooth their dividend payments. See Marsh, Merton (1986), for instance.

[30]See LeRoy (1989).

Hence, in the most general case only the value of a mutual fund containing the asset in question, and accumulating the cash flow connected with it, can be regarded as following a martingale, and simultaneously, as being describable by the stochastic difference equation (3.4).

Therefore, we can conclude that the identification of informationally efficient capital markets, in terms of the present value relation with the martingale property of asset prices, is only valid under additionally restringent assumptions, or for alternatively defined asset price movements in terms of mutual fund values[31]. One must be aware of this fact when examining the efficiency of markets. After discussing the relevance of the fundamental part of the general solution of equation (3.4), we now turn to the economic relevance of the bubble solution.

3.1.2 Rational bubbles on capital markets

The reason why the bubble solution becomes interesting in the context of capital markets, lies in its capability for capturing some phenomena on capital markets which simple present value models do not explain. Notice that although the concept of bubbles is consistent with the notion of rationality (bubbles occur in a rational expectations framework), it contradicts the paradigm of market efficiency. In discussing the relevance of rational bubbles for explaining the behaviour of asset prices, we will first present some stylized facts on capital markets. Then, a descriptive analysis of how rational bubbles may arise, grow, and crash, will be provided based mainly on socio-psychological arguments.

The following *stylized facts* observed on capital markets can be pointed out. Notice that they are by no means comprehensive, but chosen primarily because of their relation to the concept of rational bubbles.

- The most important empirical phenomenon, which was a main inspiration in the development of theoretical and empirical work on

[31] Another alternative for establishing the equivalence at once is to claim prices plus dividends follow a martingale. See LeRoy (1982).

rational bubbles, was the observation of *excess volatility* on capital markets initiated by the seminal papers of Shiller (1979, 1981b) and LeRoy, Porter (1981). The results of their analysis appear to contradict the simple present value description of asset prices and support the idea that additional elements beyond a fundamental component may be present on capital markets which are not only random but are expected to behave systematically and appear, therefore, to be predictable in some way. These additional elements are seen to be responsible for the partly extreme deviation of actual market prices from their fundamentally justified prices[32]. The existence of rational bubbles may then represent one possibility of explaining the apparently existing excess volatility on capital markets[33].

- Unlike the perception of excess volatility, which does not necessarily imply the existence of rational bubbles at all, but may be caused by other factors like anomalies or simply a model misspecification (irrationality, time–varying risk–premia and discount factors), the phenomenon of *crashes*, observed on asset markets from time to time, may well be captured by the notion of rational bubbles. In particular, the bursting bubble derived in the preceding chapter is able to depict a crash–like behaviour[34].

---

[32]Among others see Lehmann (1990) for a recent study which rejects the efficiency of stock markets. A more extensive discussion of literature on the efficient market hypothesis and volatility tests is provided in section 4.1.

[33]There may be other sources for excess volatility. One important phenomenon contradicting the paradigm of prevailing market efficiency was the appearance of so–called anomalies such as the small–firm effect, various calendar effects (begin–of–the–year effect) or the price–earnings–ratio effect. See for further exposition LeRoy (1989) and Shiller (1992). Besides these anomalies, the fact that stockholders mostly hold their assets less than a year, tends to increase price volatility (short–terminism). See Shiller (1992). Moreover, the increasing relevance of derivative markets and lower brokerage costs may additionally have enhanced the observed excess volatility.

[34]However, it must be stressed that not only the two most popular stock market crashes within this century in 1929 and 1987 might have been caused by bubbles, but even smaller bursting bubbles may have been present in asset markets during the past decades. However, they were not identified as bubbles because of their early reversion to their fundamentally justified asset price.

- Recently, another phenomenon pointed out by DeBondt, Thaler (1985), the so-called *overreaction-to-news* effect on asset prices, has entered the academic discussion. The apparent fact that asset prices overreact to (possibly fundamental) news, is mainly based on the result of psychological tests that people tend to overweigh new information and to underweigh prior[35], thereby violating Bayes' rule of information processing, which is usually viewed as an appropriate reaction to new information. Overreaction simply means that prices are more affected by news or by a general optimism than by the fundamentals themselves[36]. Until now, two possibilities for investigating the relevance of the 'overreaction hypothesis' exist in literature. First, event studies following the lines of Fama (1970) are mainly concerned with the impact of certain events like earnings and dividend announcements and examine the reaction of stock prices as related to these events. A sharp change in prices followed by an opposite correction would then indicate the existence of such an overreaction effect[37]. Second, studies like DeBondt, Thaler (1985) test for the serial dependence of stock returns. If prices overreact, an extreme increase in prices will be followed by an extreme decrease (an effect which is usually called mean reversion) implying the predictability of returns based on past returns only. DeBondt, Thaler (1985) actually show that loser (winner) stocks systematically yield higher (lower) returns compared with average returns differently defined. They interpret the result as being due to an overreaction of asset prices to extreme bad or good news.

The overreaction effect of asset prices is more general than the bubble concept since it allows for negative deviations from the fundamental value. If the price deviation is positive, rational bubbles (in particular intrinsic bubbles when the information triggering a

[35]See, for instance, Kahneman, Tversky (1982) who first emphasized that individuals apply attention irregularly.

[36]Notice that the nature of news entering the market is irrelevant. Both fundamental or non-fundamental news may have an impact on the corresponding asset prices.

[37]Interestingly, Fama (1970) found no significant impact of such announcement effects and, therefore, states a belief in the semi-strong efficiency of capital markets.

price reaction is related to the fundamentals, or extrinsic bubbles when any other newsbreak produces a price variation) have among other approaches[38] at least the potential of providing a formal framework for the analytical description, and empirical assessment, of the overreaction–to–news hypothesis.

- Finally, it was often suspected that besides a possibly too narrow economic view of capital markets (based mainly on fundamental factors related to the corresponding assets), *psychological aspects* like herd behaviour, maniacs and panics[39], or simply the phenomenon of self–fulfilling expectations[40], are relevant for determining prices[41]. The concept of rational bubbles then enables us to stick to the assumption of rational expectations, so attractive in theoretical and empirical economics, and simultaneously allows for the examination of social–psychological aspects on capital markets[42].

After justifying the consideration of rational bubbles through their potential in explaining at least part of some stylized facts in the context of capital markets, a description of how rational bubbles may emerge, grow and possibly crash – hence the entire *bubble cycle* – is useful. The description of the bubble cycle presented here relies mainly on the notion of bursting bubbles, however, it may at least partly apply to intrinsic and extrinsic bubbles, as well, especially when describing the start and growth of rational bubbles. Moreover, the arguments presented here are of a rather verbal than formal character while the formal basis was laid in the preceding chapter.

In characterizing the timing of rational bubbles and their underlying driving factors, the following aspects of a bubble cycle can be distinguished:

---

[38]For a deeper description of alternative models and tests for the overreaction effect see Meyer (1994).

[39]See Kindleberger (1978).

[40]Self–fulfilling rumors have long been argued. Keynes (1936), for instance, compares the stock market with a beauty contest where judges guess the winner of the contest not by stating their own opinion but by anticipating 'what average opinion expects average opinion to be'.

[41]See, for instance, Arrow (1982), Shiller (1984) or Summers (1986) among others.

[42]Notice, however, that the 'fad approach' described below represents an alternative for introducing psychological aspects by relaxing the rationality assumption.

- An often cited precondition for the occurrence of rational bubbles is a situation where market participants can be characterized as generally optimistic[43]. This overall *optimism* may be justified by fundamental reasons, i.e. positive economic prospects of both the entire economy, or of single firms, or, simply, by psychological aspects such as long periods of side–moving stock markets where a general feeling of 'something must happen now' arises.

- The actual appearance of rational bubbles can then be induced by different factors. A purely random event may be responsible for a fundamentally unjustified increase in asset prices or the occurrence of news, with or without fundamental relevance, may *trigger* the start of a rational bubble.

  For instance, the buy–orders of a small number of very large investors, which act simultaneously only by chance, may be seen as a purely random event causing the emergence of rational bubbles. Another possible explanation for the rise of rational bubbles simply lies in the phenomenon that if prices are expected tomorrow to be higher than today, and if this belief is shared by a group large enough to move the market, asset prices will actually increase with the force of the buy–orders of these investors (this phenomenon is usually called the self–fulfilling expectation effect)[44]. Notice that self–fulfilling expectations do not necessarily contradict the rationality assumption. If prices are expected to rise tomorrow, it is indeed rational to buy the stock today at lower price and realize a capital gain tomorrow[45]. These sources of purely random events inducing the existence of rational bubbles may best be captured by the notion of Markovian bubbles.

---

[43]See Shiller (1989, 1992).

[44]See Azariades (1981) for the notion of self–fulfilling expectations and their impact on intertemporal equilibria.

[45]This pattern reminds us of 'chain' letters which show a behaviour similar to rational bubbles. The participation in chain letters is often explained by the human tendency to gamble, which is inherently connected with the thrill of possible gains or losses. This bridges over to the fad model described below where these (irrational) psychological aspects in financial market analysis are considered more explicitly.

If some news, which are fundamentally relevant, like actual deficits of the state budget or of the trade balance, exchange rates or price level changes on the macro level, earnings or dividends annoucements of firms on the micro level, enter the market, the difficulty immediately arises as to how the new fundamental value should truly be calculated. This uncertainty concerning the correct fundamental value often implies rather an overevaluation than an underevaluation of stock prices establishing the phenomenon of overreaction of asset prices to positive fundamental newsbreaks[46]. This effect may best be described by the notion of intrinsic bubbles that include an additional non–linear fundamental term covering the theoretical overreaction effect.

Non–fundamental information like the declaration of mergers & acquisitions[47], political events, or even any other economically irrelevant news may have an impact on asset prices. These effects are usually associated with so–called focal points or sunspots. Focal points, or sunspots, display the idea that newsbreaks, although fundamentally irrelevant, may influence asset prices only because people think they have an impact. In this context extrinsic bubbles may well embody the concepts of focal points or sunspots on capital markets.

- Once a bubble has come into being, its growth may be accelerated by both social and technical factors. Notice that these *accelerating factors*, usually occurring on capital markets, may well be pictured in our model by the exponential growth property of rational bubbles[48].

  The first amplifying factors concern *social aspects*. It is often mentioned in sociological literature[49] that people's behaviour strongly

[46]See the presentation of the overreaction effect above.

[47]In the context of mergers & aquisitions a further effect is worth mentioning. Hostile take–overs are likely to increase asset prices since stockholders must be induced to sell the stocks owned. This may be anticipated and exploited by informed traders.

[48]However, price increases do not necessarily appear. Sometimes a small overevaluation of stocks will prevent investors from buying that stock, sometimes the observed price increase leads to higher investment. The reaction of market agents in face of price increases differs from time of time. There is no automatism.

[49]See Shiller (1984, 1988a) for an exposition of sociological aspects in economics.

depends upon what others do. This is explained by the fact that belonging to a group stabilizes people's psychological balance, whereas isolating from a group requires a trait of self-confidence. This may give reason for the rapid spread and adaption of information among market participants[50]. Further support may be found in the fact that some investors, due to their ability to drive the market up or down, are under special intensive inspection of all market participants in order to adopt the behaviour of those investors in time.

Moreover, this grouping behaviour may also lead to a so-called magnification of risk attitudes among market agents[51]. A magnification effect describes a situation where a pre-group risky attitude is amplified by joining a group in the sense that an individual is then willing to take more risk than before. This may partly be explained by people's tendency to give at least a part of their responsibility to the group. This magnification effect, translated into capital markets, may further amplify the market trend significantly since people tend to act more excessively when grouped together.

This dependence of individual's behaviour on the reaction of other individuals, the so-called 'herd behaviour' may mainly be induced by the following two 'ways of contagion' which serve as transmission channels on capital markets.

(a) Popular opinions concerning the market are often transmitted by oral communication. Traders know each other and talk to each other in order to get an idea of what is going on on the market. The attractiveness of this important, more direct source of information spread lies in its rapid diffusion and its low cost. The increasing use of all types of media additionally speeds up the diffusion of information and opinions. However, a certain probability may remain that some information is not available to all traders, but exclusively to only a few market participants who possibly exploit their additional

---

[50]The general epidemic model appears to be an appropriate tool for representing the spread of information among the market participants and their relevance for determining the movement of prices formally. Shiller (1989) proposes such an approach. For further description see the literature cited there.

[51]See Shiller (1984) and the literature cited there.

knowledge[52]. That is the reason why the second source of information becomes relevant.

(b) Information, possibly inducing the apparent herd behaviour on asset markets, may be transmitted more indirectly by prices themselves: Prices are seen to contain information not directly observable on the market; they serve as signals, especially when fundamentals are uncertain, and possibly indicate the existence of more informed traders[53]. Seemingly less informed traders observe other traders buying although this behaviour appears to be fundamentally unjustified. This is interpreted as a lack of information and induces further bids. The less informed traders follow the trend in accordance with a 'trend is your friend'–philosophy[54]. Hence, the argument simply states that price changes are signals conveying information from informed traders to uninformed. A price increase may then be interpreted as unrevealed positive prospects and the trend will be amplified by a corresponding reaction of seemingly uninformed traders[55].

The second factor which tends to increase the growth of rational bub-

---

[52]The phenomenon called 'window dressing' may serve as an example of this insider information. Window dressing means selling stocks that have done badly in the past so that they do not remain as reminders of past errors. If somebody knows that large amounts of the corresponding stock will be sold, he may anticipate and exploit the following price decrease.

[53]This was first formally stated by Grossman, Stiglitz (1980). The effect of more general herd behaviour aspects on capital markets related to the signalling function of asset prices is examined by Topol (1991) and Lux (1995).

[54]Although everyone engaged on capital markets must admit that excessive price increases, beyond any fundamental justification, will reverse in the distant future, everyone follows this strategy almost surely. This may at least partly be explained by the fact that portfolio manager are highly focused on the short run since their portfolio performance is evaluated in equally distinct short periods. This short–terminism may be an important source for speculation, as Shiller (1992) pointed out. Furthermore, based on the results of runs–test, this trend philosophy is inspired by the observation that price increases are more likely to be followed by further increases than by a decrease. For a description of such run–tests which examine these effects see Blanchard, Watson (1982), for instance.

[55]The existence of so–called positive feedback strategies may also be seen as a source of observed herd behaviour. For a discussion see below in the context of alternatives to rational expectation assumption.

bles stresses the relevance of *technical aspects.* Technical aspects are closely related to technical analysis conducted by chartists. Market agents are more solely concerned with the movement of market prices than with the underlying fundamentals since they believe that markets do not represent the fundamentals at all. Among other concepts, the 200–day moving average, or other long term trend lines, serving as technical indicators for buy–and–sell decisions tend to increase the movement of prices along the trend. Other trend channel concepts exist rather leading to an extrapolation of the movement of prices than to a stabilization. The general appeal of the chart analysis lies in the fact that market participants are used to thinking in trends. They extrapolate the movement of prices rather than expect a trend revision. It is this behaviour precisely that is aimed to be captured by the notion of technical aspects in the analysis of the bubble cycle here.

- After a bubble experiences a period of extensive growth, as described above, a *rebound of prices* is often expected. This is mainly initiated by the observation of declining price increases. Like a pendula in an up–swing, the bigger the bubble becomes, price expansion looses its power. This may lead to the conclusion that the price growth will come to an end. Simultaneously, a strong feeling of overevaluation of the market coincides with the perception of price increases slowing down. This feeling of an overevaluated market often precedes a sharp decline in prices. What 'overevaluation' actually means depends on the risk–aversion of the market participants. Hence, market participants perceive very sensitively all determinants (fundamental and non–fundamental) which are supposed to influence the behaviour of asset prices when these indicators are observed.

- In this situation of great sensitivity the same sources already presented above serve as a *trigger* for a crash. Arbitrary, random events like simultaneous selling of large amounts of stocks conducted by large traders, for instance, or news (fundamentally relevant or irrelevant) – now, however, negative – may induce prices to reverse.

- After once rebounded, the price drop may be accelerated by the analogous pattern of *social and technical factors* described above, now working in the opposite direction. However, an additional source of trend magnification is important in the case of extreme price declines, i.e. the effect of so–called institutional selling[56]. Institutional selling usually implies stop–loss rules adopted by investors in order to limit possible capital losses. When prices strongly decline, automated selling, due to stop–loss rules, tends to further increase the downward trend rather than to stabilize it.

Table 3.1 pictures the most relevant aspects of the bubble cycle described above and serves simultaneously as a summary of this subsection.

Table 3.1: The pattern of a bubble cycle

<table>
<tr><th colspan="4">The occurrence and growth of rational bubbles</th></tr>
<tr><td>Situation</td><td colspan="3">Feeling of optimism</td></tr>
<tr><td rowspan="3">Trigger</td><td rowspan="2">Random event</td><td colspan="2">Positive newsbreak</td></tr>
<tr><td>Fundamental news</td><td>Non-fundamental news</td></tr>
<tr><td>Markovian bubble</td><td>Intrinsic bubble</td><td>Extrinsic bubble</td></tr>
<tr><td>Amplification</td><td colspan="3">• Social aspects (herd behaviour)
• Technical aspects (trend behaviour)</td></tr>
<tr><th colspan="4">The rebound of rational bubbles</th></tr>
<tr><td>Situation</td><td colspan="3">Feeling of overevaluation of the market</td></tr>
<tr><td rowspan="2">Trigger</td><td rowspan="2">Random event</td><td colspan="2">Negative newsbreak</td></tr>
<tr><td>Fundamental news</td><td>Non-fundamental news</td></tr>
<tr><td>Amplification</td><td colspan="3">• Social aspects (herd behaviour)
• Technical aspects (trend behaviour, stop–loss rules)</td></tr>
</table>

As already mentioned earlier, the empirical fact of excess volatility, contradicting the validity of simple present value models for determining asset prices, may be explained by introducing rational bubbles which tend to increase the second moment of asset prices due to their explosive nature. However, as shown above, present value models and simultaneously the

[56] Another expression is portfolio insurance while computer or program trading also consider institutional buying. All these concepts involve buying stocks when prices rise and selling stocks when prices fall.

concept of rational bubbles rely on several partly controversial assumptions. A violation of these assumptions may also induce asset prices to fluctuate more than fundamentally justified. Hence, in the following subsection we will discuss the relevance of the assumption structure more extensively.

3.1.3 Economic caveats

At the end of chapter 2 restrictions under which rational bubbles can theoretically occur have been examined. General equilibrium arguments, the necessity of positive rational bubbles and the fact that bubbles have to be present from the first day of trading and can only burst incompletely, are derived from the theoretical framework of rational bubbles although strongly dependent on the model considered. Moreover, we have specified the precise structure of assumptions related to equation (3.1). These assumptions underly both the fundamental solution and simultaneously the bubble solution. A possible violation of these assumptions may rule out rational bubbles and may additionally explain the observed excess volatility on capital markets.

Three main assumptions have been worked out in the preceding subsection, first the assumption of rational expectations connected with a homogenous spread of information, then the assumption of risk–neutral market participants and finally a constant interest rate. All three assumptions will be discussed here in greater detail.

Alternatives to rational expectations

Unlike the assumed rationality of market participants, Shiller (1984) developed a model for explaining asset pricing which integrates the idea of *fads and fashions* influencing the movement of prices. A fad represents an extension of the present value model in the sense that a so–called fad variable F_t is included additionally. Camerer (1989) formalizes such a fad model as

$$p_t = \sum_{i=0}^{\infty} \frac{1}{(1+r)^i} E[d_{t+i} \mid I_t] + F_t$$

The fad variable is often assumed to follow the following stochastic process $F_t = cF_{t-1} + \epsilon_t$. Here, c represents an arbitrary constant denoting the impact of the lagged fad variable. c may be interpreted as the speed of convergence or decay of a fad[57]. ϵ_t stands for an iid–error term. For $c = 1 + r$ a fad equals a rational bubble. However, due to the arbitrarity of c, fads reveal to be more general than rational bubbles. Notice, that for $c \neq 1+r$ we leave the framework of rational expectations models indicating fads as an alternative to the assumption of rationally acting individuals[58].

Fad variables, in general, capture the idea that socio–psychological aspects may influence the movement of asset prices. Shiller (1984), for instance, emphasizes the effect that an increased interest in the stock market observed during the 1950s may have induced the booming stock market in that period. It simply became in fashion to participate in the stock market which may possibly express a general increase in the inclination to gamble – at least in the case of private investors. Moreover, the overall knowledge of how stock markets function rose significantly and institutional stock brokers appeared on the market for the first time governing large amounts of money owned by institutional investors. All these factors may have led the stock market to move upwards without any fundamental justification, and provides an interesting explanation of the boom during the 1950s[59].

More formally, the fad variable F_t may respond to certain variables like (lagged) returns, $F_t(r_t)$ or may depend on the fundamentals $F_t(d_t)$ due to changing utility from holding assets over time. Or a fad variable may represent the general notion of market psychology captured by varying expectations concerning market fundamentals $F_t(E[d_{t+i} \mid I_t])$[60]. While the first two sources can be viewed as fundamentally reasonable factors, the third component stands for a genuine fad or fashion variable. However, it

[57]For $c = 0$ the fad disappears immediately, for $c = 1$ the fad is very slowly mean reverting.

[58]The formalization of a fads model by Camerer (1989) represents only one possibility. Shiller (1984) provides an alternative description.

[59]Another phenomenon which may be due to the existence of a fad, is the excessive enthusiasm for trading one special asset while others are ignored without any fundamental explanation.

[60]See Shiller (1984) or Camerer (1989).

remains difficult to recover the decisive factors which may represent the notion of fads or fashion in this model. While this concept is even more general than the concept of rational bubbles, its high degree of generality implies the disadvantage of low tractability – especially concerning the derivation of testable representations of the fad model.

Another alternative to the assumption of rational expectations was introduced by Black (1986). He stresses the relevance of so–called *noise traders* on capital markets. Noise traders mainly rely on possibly irrational expectations and, therefore, induce a systematic bias to market prices with respect to their fundamental price. They are often seen as driven by their (irrational) sentiments or as reacting sensitively to pseudo–signals thought as possibly conveying information like advices of brokers or financial gurus[61]. Notice that a certain parallel to the concept of extrinsic bubbles appears. One behaviour often ascribed to noise traders is the use of inflexible trading rules like the positive feedback strategy[62]. According to this strategy, it is recommended to stick to the trend, at least in the short run, although it is almost certain that prices will reverse to their fundamentals in the long run. Hence, if the market moves up, noise traders buy stocks and they sell them if the market moves down.

Besides noise traders, so–called informed traders (smart–money traders) are considered in these models. These informed traders are assumed to behave rationally. Here the difference to the fad approach becomes obvious. While the above alternative still relies on the assumption of a homogenous spread of information, here two groups with heterogenous information structure are considered explicitly[63]. DeLong, et al. (1990b) present a model where the positive feedback strategy of naive traders is exploited by informed traders by driving up the market through strategic buying. When positive feedback traders observe rising prices, they enter the market, thereby further increasing asset prices. Informed traders then

---

[61]See Shleifer, Summers (1990). Their paper simultaneously provides a good overview of the noise trader approach and related issues. See also De Long et al. (1990a).

[62]Notice that in the description of the bubble cycle this effect has already been mentioned. We called it the 'follow the trend'–strategy.

[63]Notice that the information structure also implies the market model itself applied for the expectation formation process.

sell their assets at high prices in order to realize capital gains. The following decrease in asset prices will additionally be amplified by the selling reaction of the positive feedback traders. Hence, the interaction of informed and noise traders induces the market price to artificially rise up and break down, only due to the positive feedback strategy pursued by noise traders.

In evaluating the noise trader approach, one must point out that the existence of noise traders and especially the related positive feedback trading strategy may indeed have some explanatory power[64]. The question that arises is only whether the influence is large enough to explain the extreme deviation of actual prices from their fundamental counterparts. Further empirical research is required in this area in order to reach a final assessment of a model with heterogenous market participants.

Alternatives to risk–neutrality and constant discount rates

As already noted earlier, decisive assumptions for deriving equation (3.1) concern the risk attitude of market agents and the variation of discount rates. A natural extension of the present value model is to include a (time–varying) risk premium c_t and a variable discount rate r_t. The corresponding present value relation has then the following form

$$p_t = \sum_{i=0}^{\infty} E\left[\frac{d_{t+i}}{\prod_{j=0}^{i}(1 + r_{t+j} + c_{t+j})} \mid I_t\right] \tag{3.12}$$

Principally, three different possibilities exist in order to improve the simple present value relation.

Including a constant risk premium while r_t remains constant

Risk aversion may be incorporated by admitting a constant risk premium which drives the interest rate above the interest rate on a riskless asset. This extension has first been emphasized by Samuelson (1965)[65]. More

[64] In particular, the existence of noise traders may explain why actual trade on capital markets takes place. If all traders are efficiently informed, prices reflect all information, and no incentives for buying and selling exist.

[65] See also Blanchard, Watson (1982) in the context of rational bubbles on the gold market and Frenkel, Froot (1986) in the context of exchanges rates.

precisely, the general capital asset pricing model (CAPM), which is a standard equilibrium framework in finance, emphasizes the dependence of the return to a single asset from their covariance with an efficient market portfolio. The so-called β-factor, which includes the correlation between single stock return and market return, can then be interpreted as a risk-premium associated with the stock, respectively.

However, this extension is thought of as being too simple to explain a great part of the observed excess volatility on asset markets. Hence, no empirical investigations have been undertaken in order to test the hypothesis of a constant risk premium.

Including a time-varying risk premium while r_t remains constant

A more convincing approach uses a time-varying risk premium. Here the risk premium is related to a certain variance component which is thought to have an impact on the movement of stock prices and for which individuals should be compensated. Pindyck (1984), for instance, argues that the variance in gross marginal returns on capital induced the observed decline in stock prices during the 1970s based on a continuous time model. By decomposing the different effects considered, he shows that the effect of inflation and the average growth rate of gross marginal returns have a greater impact, for a relative risk aversion parameter $\delta < 2$, while for $\delta \gg 2$ the effect of a changing variance of the gross marginal return dominates.

While Pindyck (1984) stresses the influence of risk components on stock prices (in particular, the variance of the gross marginal return on capital) Poterba, Summers (1986) dispute this view by assessing the impact of time-varying risk premia on the present value relation. They describe the risk premium as a stochastic process depending on the variance of the equity return σ_t^2 with δ as a proportional factor, i.e. $c_t = \delta\sigma_t^2$ and found no evidence for a long lasting effect of time-varying risk premia on stock prices[66].

Including a time-varying discount rate

Many extensions of simple present value models simply assume a non-

[66]Notice that both approaches relate the risk perception to a variance component while the risk aversion parameter itself remains constant.

constant discount rate. The source of this variability differs from model to model. One of the most influencing approaches in this context is that of Grossman, Shiller (1981). They refer to the microfoundation of the present value model by explaining the variation in discount rates through changes in consumption. Recall the CRRA–utility function discussed above. According to the first order condition, the following equation was obtained

$$\frac{1}{1+r_t} = \vartheta \left(\frac{E[u'(c_{t+1})]}{u'(c_t)} \right)^{\delta}$$

Hence, discount rates vary because of variable consumption. Grossman, Shiller (1981) use precisely this relation for the calculation of the discount rate and, therefore, for the present value relation by computing the present value for $\delta = 0$ and $\delta = 4$. The former specification represents the case of risk–neutrality while the latter considers risk–averse individuals. It turned out that by assuming risk–aversion they are able to describe a part of the apparent excess volatility until the 1950s on the US–stock market. Thereafter, the variability in stock prices remains unexplainable by this model. Hence, Grossman, Shiller (1981) conclude that no decisive improvement can be reached through admitting time–varying discount rates[67].

A second line of research concentrates on a linearization of the present value model (3.12). Based on a previously derived variance bound, Shiller (1981b) computes the standard deviation of the average one–period rate of return which is necessary for satisfying this variance bound precisely. By using different data sets, Shiller (1981b) derives standard deviations of 4.36 % and 7.5 % exceeding clearly the actual standard deviation of the rate of returns which lies around 2 %[68]. Shiller (1981b) concludes that time–varying discount rates cannot be the source of apparent excess volatility on stock markets. West (1988b) extends this analysis by considering an

[67]The argument of Grossman, Shiller (1981) mainly relies on the comparison of two series plotted, one related to the present value model as calculated above, one related to the actual price path. See Kleidon (1986b) for a discussion of the appropriateness of this argument. West (1988b) repeats this procedure by enlarging the set of parameters ϑ and δ and the stochastic processes d_t and was able to confirm the results obtained by Grossman, Shiller (1981).

[68]Poterba, Summers (1986) and West (1988b), also using a linearized model but additionally allowing for unit roots, arrive at even higher standard deviations.

alternative variance bound and different fundamental processes. His results again confirm the conclusion drawn by Shiller (1981b).

Hence, although it appears to be commonly accepted that part of the excess volatility is due to varying discount rates, differing rates of interest provide by no means an exclusive explanation for the apparent rejection of simple present value models[69].

To conclude, when relaxing the underlying assumptions, the present value model is likely to come closer to reality. However, the improvement in explaining real world phenomena is not sufficient in describing the overall extent of the excess volatility observed on capital markets. The approach of rational bubbles, therefore, represents at least a potential improvement in determining asset prices.

3.2 Foreign exchange markets

The second example of equation (3.1) relates to foreign exchange markets, in particalur to the monetary model of exchange rate determination[70]. In this section we aim, again, to reveal the structure of the underlying assumptions necessary for deriving the stochastic difference equation under consideration[71]. Then, both solution components will be examined and

[69]Campbell, Shiller (1989) seem to, also, cast doubt on the view that variable discount rates may account for the apparent excess volatility. However, Cochrane (1992) recently presented a testing strategy which enables a decomposition of various effects relevant in determining stock price volatility. The results of his analysis do not lead to any strong rejection of present value models casting doubt on the results cited above. In particular, concerning time–varying discount rates, Cochrane (1992) concludes that 'discount rate processes exist that explain the variance of price–dividend ratios. But none of the discount rate models considered here ... are forecastable enough from price–dividend ratios to explain the variance of price–dividend ratios.' (Cochrane (1992) p 270). Therefore, the problem appears to be an open issue.

[70]See, for instance, Mussa (1976). A good overview of monetary exchange rate models gives Levich (1985).

[71]Unfortunately, the underlying assumptions of the basic monetary model of exchange rates cannot easily be derived from a possibly more convincing intertemporal optimization framework. Hodrick (1987) Chap. 2 relying on Lucas (1982) provides a rather uncomfortable approach. A complete presentation is omitted here.

evaluated by assessing their potential for explaining some stylized facts on foreign exchange markets. Finally, the alternatives to the assumptions recovered here will be discussed shortly.

The first decisive assumption concerns the system of exchange rates under which the phenomenon of rational bubbles is likely to occur, i.e. most of the models of rational bubbles on foreign exchange markets rest upon the assumption of perfect flexibility of exchange rates.

Assumption 4 Exchange rates are freely floating.

In this framework consider the Keynesian money market for two identical countries describing the demand for real money balances $m_t - p_t$, as depending on a transaction motive positively related to national income y_t, and a speculative motive negatively correlated with the interest rate of an alternative investment on the capital market i_t.

$$m_t - p_t = ay_t - bi_t \tag{3.13}$$

$$m_t^* - p_t^* = ay_t^* - bi_t^* \tag{3.14}$$

All variables except those of the interest rates are expressed in logs. Foreign variables are denoted by an asterisk. The parameters $a,\ b > 0$ represent the income elasticity and the semi–elasticity of real money demand concerning changes of the interest rate[72]. If equilibrium prevails on both money markets, real money demand equals money supply in every period.

Both money markets are connected through two arbitrage conditions, one related to the goods market and one related to the international capital market: i.e. the purchasing power parity and the (uncovered) interest rate parity both relying on the assumption of perfect mobility.

Assumption 5 Goods are perfectly mobile and both domestic and foreign goods are perfect substitutes for each other, implying the validity of the purchasing power parity. Notice that the validity of the purchasing power parity relies on perfectly flexible prices in both countries.

[72]Notice that these parameters are equal in both equations, due to the assumption of identical countries. Admitting divergent parameters does not change the outcome.

Equivalently, assets are perfectly mobile and domestic and foreign assets are perfect substitutes for each other, implying the validity of the (uncovered) interest rate parity.

Both parities can be formulated as

$$s_t = p_t - p_t^* \tag{3.15}$$

$$i_t - i_t^* = s_{t,t+1}^e - s_t \tag{3.16}$$

Equation (3.15) represents the (absolute) purchasing power parity, stating that the log–spot exchange rate s_t, quoted in terms of home currency per unit of foreign currency, equals the difference between log–domestic and log–foreign price levels, i.e. the exchange rate is a relative price of home and foreign price levels. Equation (3.16) describes the expected change in the exchange rate $s_{t,t+1}^e - s_t$ as depending on the interest rate differential $i_t - i_t^*$ between both countries. Under assumption 1 (rational expectations) the interest rate parity can be reformulated as

$$i_t - i_t^* = E[s_{t+1} \mid I_t] - s_t \tag{3.17}$$

Notice that again a homogenous spread of information assumption is often related to the assumption of rational expectation. However, this is by no means necessary for deriving bubbles within this framework. This issue will be discussed below in greater detail. Furthermore, for deriving the interest rate parity, no risk components are taken into account. Only return components play a role. Again, assumption 2 (risk–neutrality) is obtained. The assumptions derived so far represent the framework in which rational bubbles are derivable. Their alternatives will be discussed below. Simple rearrangement yields our stochastic difference equation. Equations (3.13) and (3.14) can be composed to

$$p_t - p_t^* = m_t - m_t^* - a(y_t - y_t^*) + b(i_t - i_t^*)$$

If the purchasing power parity (3.15) is valid, the exchange rate can be expressed as

$$s_t = m_t - m_t^* - a(y_t - y_t^*) + b(i_t - i_t^*)$$

Taking into account the interest rate parity (3.17), the well known stochastic difference equation is obtained

$$s_t = \frac{b}{1+b}E[s_{t+1} \mid I_t] + \frac{1}{1+b}(m_t - m_t^*) + \frac{a}{1+b}(y_t - y_t^*)$$

The equivalence to equation (3.1) can easily be established by equating $\alpha = \frac{b}{1+b}$ and $x_t = \frac{1}{1+b}(m_t - m_t^*) + \frac{a}{1+b}(y_t - y_t^*)$. In this case, the parameter α of the reduced form model depends on the structural parameter b standing for the semi–elasticity of real money balances concerning the interest rate. Under the assumption $b > 0$ the parameter α lies between zero and one. Again the forward solution is applicable. Based on the results derived in chapter 2, the forward solution has the following form

$$p_t = \sum_{i=0}^{\infty} \left(\frac{b}{1+b}\right)^i E[x_{t+i} \mid I_t] + \left(\frac{1+b}{b}\right) M_t$$

where M_t denotes a martingale and x_t is defined above. According to the fundamental component, exchange rates are determined by the expected future differentials of fundamentals (national income and real money balances) between both countries[73]. Notice that the fundamentals derived here can be influenced by fiscal and monetary policy actions which makes estimating the future realization of fundamentals extremely difficult. The bubble component again contains a martingale and grows at rate $\frac{1}{b}$, i.e. the inverse of the semi–elasticity of real money balances concerning the interest rate.

Empirical examination of the efficiency of *foreign exchange markets*, analogous to efficient capital markets, is again mainly based on two different approaches, as presented above[74]. The first line of research tests for the

---

[73] Meese, Singleton (1983) extend the description of fundamentals by simply assuming a vector of variables z_t thought as to determine the spot rate which contains interest rate or price level differentials or the current account in addition to income and money demand differentials.

[74] However, again a definition of efficient foreign exchange markets is far from being unambiguous. Some authors associate efficiency simply with the notion of rational expectation. See Gaab (1990). Alternatively, the notion of efficiency is interpreted as the lack of profitable trading rules. Runs tests try to assess the probability of different trading rules. Moreover, often differently defined no–arbitrage conditions (triangular, covered interest rate parity) are tested in order to gain evidence of the efficiency of foreign exchange markets.

martingale property of exchange rates[75] and is mainly stimulated by the seminal paper of efficient capital markets by Fama (1970). The second relies on the fundamental solution of the general solution as derived above and tests for the validity of the present value relation. Again, both concepts can be related by following the same lines as before. Whereas the first approach of testing for the efficiency of foreign exchange markets usually does not reject the efficiency hypothesis, in the sense that exchange rates can be described as following a martingale, the second, however, often denies the validity of the efficient market hypothesis. The latter phenomenon is then called excess volatility on foreign exchange markets, currently the prevailing paradigm[76].

Inspired by the popularity of rational bubbles on capital markets, this apparent excess volatility was often ascribed to the existence of *rational bubbles* capturing the same aspects as derived above in the context of capital markets (i.e. self–fulfilling expectations, social and technical aspects)[77]. However, proving the existence of rational bubbles on foreign exchange markets appears to be difficult. The evidence in favour of the presence of rational bubbles was mainly derived indirectly through the application of indirect testing strategies which will be described more precisely in section 4.1. Direct tests for exchange rate bubbles have not been conducted yet. The evidence coming from indirect tests can only be interpreted as a failure of simple present value models, however, and does not indicate the existence of bubbles directly.

---

[75]For an overview see Levich (1985, 1989) or Hodrick (1987, 1990).

[76]See, for instance, Huang (1981) and the deeper discussion of so–called volatility tests in section 4.1.

[77]The list of stylized facts in the context of foreign exchange markets is significantly shorter than in the context of capital markets. Phenomena such as extreme crashes observed on capital markets seldom appear on foreign exchange markets. One well–known historical example, which has been attributed to the existence of rational bubbles, was the apparent overevaluation of the Dollar during the 1980s followed by a Dollar–collapse in 1985. See Frenkel, Froot (1986). For a fundamental explanation of this phenomenon see Garber (1990b). Analogous to the overreaction effect on capital markets, a so–called *news dominance effect* seems to be present on foreign exchange markets describing the phenomenon that the variation in spot returns, $s_{t+1} - s_t$, is much larger than the variation in the forward premium, $s^f_{t,t+1} - s_t$. This may be explained by the occurrence of news between t and $t+1$ possibly capturable by the notion of rational bubbles.

This difficulty in testing for rational bubbles in foreign exchange markets is further exaggerated by the existence of many similar effects which have, also, the potential to explain the observed excess volatility on foreign exchange markets. These alternatives will now be discussed following the assumptions derived above.

Unlike the assumed system of perfectly floating exchanges rates, the reality on foreign exchange markets is characterized by frequent intervention of central banks. Usually a perfectly floating system prevails only between a few countries. Between all major currencies a so-called *managed floating system* is pursued by central banks aiming to hold exchange rates within a prespecified target zone through discretionary interventions[78]. These target zones may be formally installed like the European Monetary System or may be of more informal character like unofficial target zones between the Dollar, DM and Yen. The concept of target zones was first brought into discussion by Williamson, Miller (1987). A formal framework for the examination of target zones, applying modern techniques of stochastic calculus, was recently provided by Krugman (1991) and Froot, Obstfeld (1991a)[79]. These target zone models usually describe the upper and lower bands of a target zone as so-called absorption or reflection points, where exchange rates, displayed by a stochastic process, are either absorbed or reflected when reaching the border of the target zone representing a central bank's intervention on the foreign exchange rate market. Interestingly, in this formal framework of a target zone, Buiter, Pesenti (1990) are also able to derive the possibility of rational bubbles. However, testing for the presence of rational bubbles in a target zone model remains difficult, due to the complexity of the stochastic process normally considered in these models[80].

---

[78]Alternatively, a system of pegged exchange rates exists. In models with fixed exchange rates an effect can theoretically be derived that is called 'speculative attacks' describing the consequences of a balance of payments crisis for the exchange rate movement. See Flood, Garber (1984), for instance, or Agenor, Flood (1994) for a good exposition of speculative attacks. These effects show a behaviour similar to rational bubbles. However, speculative attacts only occur in a system of fixed exchange rates while rational bubbles are related to floating systems.

[79]See, also, Bertola (1994) for a good exposition of target zone models.

[80]For instance, these models are usually formulated in continuous time.

Unlike the purchasing power parity, which relies on the assumption of perfectly flexible prices, money market models exist describing prices on goods markets as adjusting more slowly than prices on capital markets. The most popular example is the *overshooting model* designed by Dornbusch (1976)[81]. In this model the stickiness of goods prices causes an overshooting of exchange rates after an exogenous shock. Exchange rates bear the entire burden of the short–term adjustment. In the long run prices again approach their equilibrium value and the overshooting effect is reduced. This overshooting effect may easily be interpreted as a rational bubble due to its similar behaviour. One must be aware of this similarity when testing for rational bubbles on foreign exchange markets. Interestingly, rational bubbles can also be derived within an overshooting framework. Miller, Weller (1990) and Pelloni (1993) theoretically show that both effects – overshooting and rational bubbles – may simultaneously occur. However, an empirical investigation of the explanatory power, together with an convincing distinction of both concepts, again appears very difficult mainly due to their great similarity[82] and, therefore, has not been carried out yet.

Concerning the rationality assumption, only few approaches exist which try to integrate concepts alternative to rational expectations. These approaches focus mainly on the homogenous spread of information assumption, usually associated with rational expectations[83]. Frankel, Froot (1986), for instance, investigating the strong appreciation of the dollar during the 1980s, construct a model with *heterogenous individuals*. Fundamentalists

---

[81]Meese (1986) proposes an alternative approach by using a model which describes deviations from the purchasing power parity as a random walk. For a similar approach see Pittis (1993).

[82]Another problem arising in the context of rational bubbles on foreign exchange markets is the apparent affinity to possible changes in fundamentals taken into account by market agents, but not observed by the econometrician. This may induce a bubble–like behaviour of exchange rates. This effect is usually called the 'Peso–effect' or simply 'process switching' and will be discussed in greater detail in subsection 4.2.3. The 'Peso–effect' refers to fluctuations in the Peso forward rate in anticipation of a devaluation that did not occur in the sample period. See Obstfeld (1987) for an investigation of the Peso–effect on the foreign exchange market.

[83]Fad models on foreign exchange markets, capturing more directly irrational behaviour or other psychological effects, do not exist in literature to date.

are described as explaining the movement of exchange rates in accordance with Dornbusch's overshooting model, chartists use a simple extrapolation technique in order to estimate the behaviour of exchange rates, while portfolio manager, as a third group, apply both models for prediction purposes by putting a time–changing weight on both models' forecasts. In their model, rational bubbles can also occur although individual information is heterogenous.

Arguments similar to those adapted in the preceding section may be pointed out concerning the risk–neutrality assumption. Again, a constant *risk–premium* derived from an international capital market model[84] may be considered as an additional component within the interest rate parity, however, the inability of explaining the data remains. The issue becomes more complex when taking into account a time–varying risk–premium. Hodrick (1987)[85] discusses this issue at length and concludes that 'none of the models are capable of the task' (p 157), i.e. to explain excess volatility on foreign exchange markets.

To conclude, rational bubbles may theoretically occur on foreign exchange markets. However, on foreign exchange markets many effects exist which interfere with the perception of rational bubbles. These alternative effects such as the overshooting–phenomenon, the Peso–effect or central bank's intervention policy in connection with the difficulty in assessing future fundamentals which often serve as political target variables, may make the evaluation of rational bubbles on foreign exchange markets extremely difficult[86]. Moreover, among the variety of exchange rate models that exist in literature, no one is commonly accepted. Since the notion of bubbles depends on a correct specification of fundamentals, which can only be determined in an economic model, the severe difficulty in empirically assuring the relevance of rational bubbles on foreign exchange markets becomes obvious.

---

[84]See, for instance, Dumas (1994).

[85]See, in particular Hodrick (1987) Chap. 5.

[86]This is one reason why we prefer to test for the existence of rational bubbles on capital markets rather than on foreign exchange markets.

3.3 Hyperinflation

The last application of equation (3.1) in economic theory is related to the hyperinflation model of Cagan (1956). Cagan characterizes the demand for real money as depending mainly on the expected rate of inflation if extreme price increases are observed. The structural form of the money market model, which represents an inflationary counterpart to that one presented in the preceding section, can be formulated as

$$m_t^d - p_t = \gamma_0 - \gamma_1(p_{t,t+1}^e - p_t) \tag{3.18}$$

$$m_t^s = z_t \tag{3.19}$$

$$m_t^d = m_t^s \tag{3.20}$$

All variables are again expressed in logs. Equation (3.18) represents the demand for real money balances to react to changes in expected rates of inflation $(p_{t,t+1}^e - p_t)$. In recent studies of periods of hyperinflation, the assumption of adaptive expectations, originally imposed by Cagan, was removed in favour of assumption 1 (rational expectations). Then, equation (3.18) becomes[87]

$$m_t^d - p_t = \gamma_0 - \gamma_1(E[p_{t+1} \mid I_t] - p_t) \tag{3.21}$$

The coefficient γ_1 is positive ($\gamma_1 > 0$) and stands for the elasticity of real money demand with respect to the expected rate of inflation. The parameter γ_0 summarizes all additional factors usually considered as influencing the demand for money such as the real interest rate or the level of income. During periods of hyperinflation these factors are assumed to be dominated by the impact of the expected rate of inflation and, therefore, are assumed to be relatively constant. Hence, a further assumption is

Assumption 6 The inflation rate is the dominating factor for money demand in periods of hyperinflation. Traditional variables usually thought of as having a certain impact on money demand are assumed to be constant.

[87] An additional error term is often included in equation (3.21) which is often described by a random walk. See, for instance, Casella (1989) and Engsted (1993b). It is omitted here for the sake of simplicity.

Due to the rational expectation assumption, the corresponding information set I_t contains all current and past values of m_t and p_t as well as all information concerning the neglected variables comprised in the parameter γ_0.

Equation (3.19) determines the money supply as an exogenous stochastic process z_t. This specification is by no means necessary. Some authors, for instance, suggest special feedback rules of money supply on the actual price level[88]. While the theoretical derivability of rational bubbles appears to be independent of the underlying causality structure, the assumption of endogeneity or exogeneity of money supply becomes an important issue in testing for the rational bubble hypothesis in the sense that the outcome differs depending on the assumed characterization of causality[89]. Consequently, a test design aiming to detect rational bubbles in periods of hyperinflation must bear in mind this sensitivity.

Finally, equation (3.20) characterizes the equilibrium prevailing on the money market at all t. The structural form of the model can then easily be reduced to our well-known stochastic difference equation by simple rearrangement.

(3.22)
$$p_t = \frac{\gamma_1}{1+\gamma_1} E[p_{t+1} \mid I_t] + \frac{z_t - \gamma_0}{1+\gamma_1}$$

The equivalence to equation (3.1) can be established by identifying $\alpha = \frac{\gamma_1}{1+\gamma_1}$ and $x_t = \frac{z_t - \gamma_0}{1+\gamma_1}$. Notice that $\alpha < 1$ since $\gamma_1 > 0$. Therefore, as fundamental solution component the forward solution applies.

Again, a microfoundation can be provided[90] since equation (3.22) is shown to be derivable from an intertemporal optimization model in which indi-

---

[88] For instance, money supply is described as $m_t = \rho_0 + \rho_1 p_{t-1} + \epsilon_t$ with ρ_0, ρ_1 being parameters and ϵ_t white noise. Notice that now the corresponding stochastic difference equation becomes univariate since the exogenous stochastic process vanishes in the resulting stochastic difference equation. See McCallum (1983) or Evans (1986).

[89] Flood, Garber (1980), Flood et al. (1984), Burmeister, Wall (1982) conduct tests for hyperinflationary bubbles under the assumption of exogenous money supply whereas Burmeister, Wall (1987), Casella (1989), and Engsted (1993b) consider additionally an explicit feedback mechanism. For a more detailed discussion of their tests and the related results see chapter 4.

[90] Among others see Brock (1974, 1975), Kingston (1982), Obstfeld, Rogoff (1983, 1986) and Gray (1984). All models optimize over an infinite horizon, often associated with the perfect foresight assumption, simplifying the rational expectation framework.

viduals maximize their expected utility depending additive separably on current and future consumption and where ϑ again denotes individual's time preference.

$$\max_{c_{t+i}} \sum_{i=0}^{\infty} \vartheta^i E[u(c_{t+i}) \mid I_t]$$

Assuming that the consumption good is not storable, money is the only asset and serves as an instrument of intertemporal transfer of purchasing power, substituting other saving opportunities. Given an exogenous stream of income w_t, the individual maxizimes his utility under the following budget restriction valid for all periods t

$$c_t = w_t - \frac{M_t^d}{P_t} + \frac{M_{t-1}^d}{P_t} \tag{3.23}$$

where M_t^d denotes the nominal money demand of the individual in t, and P_t stands for the price level corresponding to period t. Substitution of equation (3.23), adequately defined for two successive periods t and $t+1$, into the utility function, and differentiation with respect to M_t^d, yields the following first order condition (Euler-equation)

$$\begin{aligned} -u'(c_t)\frac{1}{P_t} + \vartheta E\left[u'(c_{t+1})\frac{1}{P_{t+1}} \mid I_t\right] &= 0 \\ \Leftrightarrow \quad \vartheta \frac{E[u'(c_{t+1}) \mid I_t]}{u'(c_t)} &= \frac{E[P_{t+1} \mid I_t]}{P_t} \end{aligned} \tag{3.24}$$

It can easily be seen that optimality prevails if the marginal utility, gained from consuming one unit of real money in t, equals the marginal utility of that unit consumed in $t+1$ times the time preference parameter ϑ. The first order condition (3.24) determines implicitly the optimal consumption path and, therefore, the optimal demand for real money balances. Both depend on the ratio of expected future and current price levels or equivalently on

---

Some are formulated in discrete time, some in continuous time. Some assume the endogeneity of the production sector, some rely on an exogenous production side. Many models, however, include money in the utility function describing the effect that holding money reduces transaction costs and, therefore, increases utility. Notice that here money mainly provides the only storage possibility and serves, therefore, as a savings opportunity, not as a mean for reducing transaction costs. See also footnote 91.

the rate of inflation, the former positively, the latter negatively[91]. This implicit money demand function can be expressed as

$$\frac{M_t^d}{P_t} = L\left(\frac{E[P_{t+1} \mid I_t]}{P_t}\right) \quad \text{with} \quad L' < 0$$

After taking logs, and under the assumption of exogenous money supply, the following monetary equilibrium condition can be derived

$$m_t - p_t = \gamma_0 - \gamma_1(E[p_{t+1} \mid I_t] - p_t)$$

where lower letters indicate log-variables. The equivalence to equation (3.22) is obvious.

After establishing the stochastic difference equation (3.22) from an intertemporal optimization framework, consider the general solution to equation (3.22) as already derived in the previous chapter. Notice that because of $\alpha < 1$ the forward solution is valid. Hence, the general solution takes the following form

$$(3.25) \quad p_t = \frac{1}{1+\gamma_1}\sum_{i=0}^{\infty}\left(\frac{\gamma_1}{1+\gamma_1}\right)^i \frac{E[z_{t+i} \mid I_t] - \gamma_0}{1+\gamma_1} + \left(\frac{1+\gamma_1}{\gamma_1}\right)^t M_t$$

Neglecting the bubble component for a moment which is usually done by imposing an appropriate transversality condition, the fundamental solution states that current (log–) price level p_t is fully determined by the expected stream of future (log–) real balances. This solution may be interpreted in two different ways depending on the direction of causality: First, the fundamental solution of equation (3.25) emphasizes the relevance of money supply as a dominant source for inflationary tendencies. In this model, high increases of money supply are exclusively responsible for (hyper–)inflation. This is not an unexpected result since a monetary approach to hyperinflation is considered here. Second, the fundamental solution implies an interesting policy advice: Other things comprised in γ_0 being equal, a

[91] Notice that although money represents a savings function here, the reaction on price changes is unambiguous: increasing inflation lowers future consumption, whereas current consumption is extended as long as $\Delta M_{t+1} \geq 0$ holds. This contradicts Blanchard, Fischer (1989) p 217 who state that the obtained elasticity of money demand on inflation could vary from $-\infty$ to $+\infty$.

monetary rule, to which the central bank is credibly committed, will stabilize price fluctuations and prevent inflation. For economies facing a period of hyperinflation the fundamental solution component of equation (3.25) recommends that controlling the money supply will dampen the extreme price variations.

However, it turns out that by no means, also in hyperinflation periods, money supply is the only factor which drives prices up. It is expected that some of the factors already mentioned above, psychological aspects like herd–behaviour or a sort of self–fulfilling expectations, contribute to hyperinflations. This may justify the consideration of rational bubbles in this setting since rational bubbles are able to exactly capture these psychological effects. A serious consequence would be that if the existence of rational bubbles during hyperinflation periods can be proven, a monetary rule which only controls the process of money supply, is ineffective in reestablishing stable prices.

Although the economic application of the bubble concept, at least during hyperinflations, appears to be reasonable[92], most of the papers concerned with monetary equilibria try to rule out these divergent equilibrium paths[93]. Two alternative equilibrium paths are usually considered: an imploding price path, describing a situation of decreasing prices, and an exploding price path, describing inflationary situations[94]. The former path is associated with a negative bubble, due to a negative initial value of the martingale, $M_0 < 0$, the latter is related to our common positive bubbles.

[92]Empirical investigations concerning the presence of rational inflationary bubbles was first carried out by Flood, Garber (1980) followed by Flood, Garber, Scott (1984) in the context of German hyperinflation during the 1920s. For a discussion of their results and related research see section 4.2.1.

[93]Notice that the discussion of arguments, in order to rule out rational bubbles for theoretical reasons, undertaken at the end of the last chapter rely precisely on these arguments presented here. Notice further that the arguments presented there are only derived in a model which describes the money market. However, money can be viewed as an asset with a negative return equal to the rate of inflation. Hence, the results applied here are also valid for more general asset models.

[94]Notice that both situations correspond to a situation where the money growth rate is converse to the rate of inflation which can be verified from equation (3.25). See, for instance, Kingston (1982).

Gray (1984), based mainly on Brock (1974, 1975), examines both equilibrium paths in a continuous time, infinite horizon setting. For deflationary price paths (imploding bubbles) Gray (1984) shows that a profitable arbitrage opportunity exists which raises utility through converting real balances in consumption, while simultaneously the Euler–equation holds. By imposing an adequately defined transversality condition, which describes this arbitrage argument, implosive price (explosive monetary paths) paths can be ruled out[95].

If explosive price paths are considered, Obstfeld, Rogoff (1983) and Gray (1984) provide further conditions under which part of these equilibrium paths can be excluded. They stress the relevance of government backing of the currency, which prevents some explosive price paths, and a certain restriction on individuals' preferences, postulating that if prices move on a (special) explosive path, utility gained from holding real balances, does not go to minus infinity if money holdings approach zero. However, these conditions are not able to exclude all explosive price paths. The existence of rational inflationary bubbles still remains possible.

To conclude, rational bubbles in models of hyperinflation are easily derived. Additionally, a microfoundation was provided. Further arguments for ruling out rational bubbles in a monetary setting were presented. However, it turns out that at least part of explosive price paths associated with explosive rational bubbles continue to be obtainable.

[95]See also Obstfeld, Rogoff (1986) for a more formal argument.

Chapter 4

On Testing for Rational Bubbles

In chapter 2 we derived a solution to a linear stochastic difference equation consisting of a particular part, representing the present value of future expected fundamentals, and a bubble part, containing an arbitrary martingale. Despite its theoretical possibility, the existence of rational bubbles remains a question which can only be answered empirically. The purpose of this chapter is to present an overview of econometric methods applied thus far in order to give some evidence of the empirical relevance of rational bubbles[1].

In general, two types of tests for the existence of rational bubbles can be distinguished: *Indirect tests* do not specify the bubble process explicitly whereas *direct tests* assume a well defined bubble process. The empirical results, related to the test procedures considered, are summarized after presenting the method.

4.1 Indirect tests

The general difficulty in dealing with rational bubbles consists in the multiplicity of the general solution to equation (2.4) caused by the existence of

[1] A first overview of testing procedures, and their results, is provided by Flood, Hodrick (1990); in the context of exchange rates see Gaab (1990).

the arbitrary martingale component. This multiplicity makes testing for the existence of rational bubbles, on the basis of the entire general solution, more complicated than testing for rational bubbles merely on the basis of the particular solution. Thus, indirect tests try to avoid the problem of choosing one concrete bubble process, among the infinite set of possible processes, through examining the validity of the particular solution alone, or by other adequately formulated null hypotheses. A rejection of the null hypothesis of no rational bubbles may then be implicitly interpreted as a hint for the existence of rational bubbles.

Following the seminal papers on the theoretical existence of rational bubbles at the beginning of the 1980s, a variety of indirect test procedures were developed. They differ in their direction of interest: Various approaches, in particular first variance bounds tests, try to examine exclusively the validity of the particular solution (i.e. of the present value relation). An intended rejection of the null hypothesis is often not further explained. Other procedures, in particular integration, cointegration and specification tests, focus more directly on the existence of rational bubbles. In the following we will present the most popular indirect test procedures, i.e. various types of variance bounds tests, specification tests and integration or cointegration tests, in greater detail.

4.1.1 Variance bounds tests

Variance bounds tests[2] originally arose in the context of the hypothesis of efficient markets, implying the validity of present value models, as presented in the previous chapter. They derive bounds on the second moment of the endogenous variable y under the null hypothesis of prevailing market efficiency, in the sense that if the present value relation provides a correct description of the reality, then the endogenous variable y, represented by a moving average consisting of future expected values of the fundamentals, should vary less than a corresponding perfect foresight endogenous variable y^*, related to the actual stochastic process of the fundamentals.

[2]Variance bounds tests are sometimes called 'volatility tests' or 'tests of excess volatility', as well. For a first overview of variance bounds tests see LeRoy (1989). A comprehensive discussion is provided by Shiller (1988c).

Empirically, however, the endogenous variable y (stock prices, for instance) appears to 'move too much to be justified by subsequent changes in' fundamentals, as Shiller (1981b) pointed out for the case of efficient capital markets. From there comes the idea of variance bounds tests: since rational bubbles tend to increase the second moment of the endogenous variable y, variance bounds tests may also serve as tests for the existence of rational bubbles. Thus, they can be thought of as simultaneously testing for the hypothesis of market efficiency, usually implying the validity of a present value model, and the existence of rational bubbles, in the sense that if market efficiency does not prevail, indicated by a violation of the variance bounds, rational bubbles may exist[3].

The earliest versions of variance bounds tests were developed independently by LeRoy, Porter (1981) and Shiller (1979, 1981b) and were frequently applied in economic studies because of their simplicity in connection with the serious results that usually contradict the paradigm of market efficiency. Following LeRoy, Porter (1981) two lines of variance bounds tests can be distinguished: First, on the basis of the present value relation, inequalities concerning the second moment of the endogenous stochastic process $\{y_t, t \in \mathcal{T}\}$ and the second moment of an ex-post rational endogenous stochastic process $\{y_t^*, t \in \mathcal{T}\}$ are derived. It can be shown that the latter variance provides an upper bound for the former variance (original variance bounds tests). Second, by including rational prediction errors, equalities concerning the average deviation of the endogenous stochastic process $\{y_t, t \in \mathcal{T}\}$ and the ex-post rational endogenous stochastic process $\{y_t^*, t \in \mathcal{T}\}$ from their mean may be tested alternatively (orthogonality tests).

Turning to *original variance bounds tests*, consider as a starting point the present value representation of the fundamental solution derived above as a forward solution of equation (2.4)[4]

$$y_t = \sum_{i=0}^{\infty} \alpha^i E[x_{t+i} \mid I_t]$$

[3]The presence of rational bubbles, however, provides only one possible explanation for the rejection of variance bounds.

[4]Recall that in most of the economic applications of equation (2.4) only the forward solution is appropriate.

The forward solution relates the endogenous variable y to an infinite sum of rationally expected future values of the exogenous variable x. Notice that because of $|\alpha| < 1$ the weights of the infinite sum decrease if $t \to \infty$. As a consequence, the effect of new information on future expected realizations of the fundamentals will only cause a damped change of the endogenous variable y to date and will be smaller the more distant the change is expected in the future.

Assuming rational expectations, we can separate the present value relation into two components, one related to an ex-post rational endogenous stochastic process (perfect foresight path), and one to a sequence of rational prediction errors $\{\epsilon_t, t \in \mathcal{T}\}$, i.e.

$$y_t = \sum_{i=0}^{\infty} \alpha^i x_{t+i} - \sum_{i=0}^{\infty} \alpha^i \epsilon_{t+i}$$

Hence, the current realization of the endogenous stochastic process $\{y_t, t \in \mathcal{T}\}$ equals the current realization of the ex-post rational endogenous stochastic process $\{y_t^*, t \in \mathcal{T}\}$ minus a sequence of rational prediction errors $\{u_t, t \in \mathcal{T}\}$[5]

$$y_t = y_t^* - u_t \tag{4.1}$$

with

$$y_t^* = \sum_{i=0}^{\infty} \alpha^i x_{t+i} \quad \text{and} \quad u_t = \sum_{i=0}^{\infty} \alpha^i \epsilon_{t+i}$$

Notice that y_t is related to the *expected* stream of fundamentals whereas y_t^* (as an ex-post rational sequence) is related to the *actual* stream of fundamentals. On the other hand, y_t represents the actual value of the endogenous variable whereas y_t^* characterizes the potential actual value of the endogenous variable if expectations were perfect, i.e. if no uncertainty prevails. As already shown in the context of rational expectations, a rational prediction error ϵ_t is uncorrelated with (or independent of) every element of the information set I_t and, therefore, also with y_t^*. Intuitively, independence of the rational prediction error means that a forecast once made by using rational expectations cannot be improved by including any

[5] Notice that the sequence of rational prediction errors starts for $i = 0$. If x_t is contained in I_t, the expectation of x_t equals its true value, and ϵ_t becomes zero. Hence, the sequence may also start with $i = 1$.

variable contained in the information set I_t. Because of independence between y_t^* and ϵ_t, it turns out that y_t is an unbiased forecast of y_t^*[6], i.e. $y_t = E[y_t^* \mid I_t]$ characterizing y_t as ex–ante rational implied by the assumption of rational expectations. Due to the independence, we are able to apply a property of statistic theory concerning the second moment of a compound variable: the variance of the sum of uncorrelated variables is the sum of their variances. Hence, from equation (4.1)

$$\text{(4.2)} \qquad \text{Var}(y_t) = \text{Var}(y_t^*) - \text{Var}(u_t)$$

Since a variance cannot be negative, we arrive at the following inequality frequently tested for in various economic fields.

$$\text{(4.3)} \qquad H_0: \quad \text{Var}(y_t) \leq \text{Var}(y_t^*)$$

This fundamental inequality of (original) variance bounds tests states that the variance of the ex–ante rational endogenous stochastic process $\{y_t, t \in \mathcal{T}\}$, i.e. the actual sequence has to be smaller or equal to the variance of the ex–post rational endogenous stochastic process $\{y_t^*, t \in \mathcal{T}\}$[7]. The rationale behind this inequality lies in the fact that since the endogenous variable y represents an unbiased forecast of the ex–post rational variable y^*, y (the forecast) should vary less than y^* (the variable being forecasted),

[6] This can be shown by taking expectations conditional on information set I_t on both sides of equation (4.1).

[7] There exist a variety of different types of variance bounds test in first papers on this issue. Inequality (4.3), for instance, was first formulated in Theorem 2 of LeRoy, Porter (1981). In addition to the upper bound in equation (4.3), LeRoy, Porter (1981) derived a lower bound for the variance of the endogenous stochastic process $\{y_t, t \in \mathcal{T}\}$, taking into account a smaller information set $H_t \subset I_t$. Moreover, Theorem 1 of LeRoy and Porter (1981) states an inequality relating the coefficient of dispersion (ratio of standard deviation and mean) of y_t and y_t^*.

Shiller (1981b) examined the same fundamental inequality as in equation (4.3). Similar inequalities formulated as upper bounds on the variance of the endogenous stochastic process were derived by Shiller (1981b) in terms of $\Delta y_t = y_t - y_{t-1}$ or ∂y_t where ∂ is the innovation operator. Both, LeRoy, Porter (1981) and Shiller (1981b) relate to efficient capital markets.

Moreover, Michener (1982) deduced variance bounds for the case of efficient capital markets and relates his results to the permanent income hypothesis. Meese, Singleton (1983) derive variance bounds for the case of efficient exchange rates.

i.e. if future realizations of fundamentals are higher (lower) than expected, y_t^* will lie above (below) y_t. If not, i.e. if y_t is more likely to lie above (below) y_t^* for high (low) values of y_t^*, y_t would be an upward (downward) biased estimator of y_t^*[8]. Notice that inequality (4.3) implies the validity of market efficiency. However, rejecting this inequality may be interpreted as markets being inefficient or that some variables, allowing for predicting future realizations of the endogenous variable, are not contained in I_t. There remains a severe ambiguity.

An alternative *orthogonality test* takes equation (4.2) as a starting point and tries to estimate each component of the equality independently. Because of the stationarity property, the infinite sum of rational prediction errors can be expressed as

$$\begin{aligned} \text{Var}(u_t) &= \text{Var}([\epsilon_t + \alpha\epsilon_{t-1} + \alpha^2\epsilon_{t-2} + \ldots]) \\ &= (1 + \alpha^2 + \alpha^4 + \ldots)\text{Var}(\epsilon_t) \\ &= \frac{1}{1-\alpha^2}\text{Var}(\epsilon_t) \end{aligned}$$

After estimating each component of equation (4.2) independently, the following null hypothesis of market efficiency, formulated as an equality,[9] is tested

$$H_0: \quad \text{Var}(y_t) = \text{Var}(y_t^*) - \text{Var}(u_t)$$

Notice that in order to carry out an orthogonality test, three point estimates have to be computed which makes orthogonality tests less attractive than pure variance bounds tests. But it can be shown that orthogonality tests have higher power than original variance bounds tests.

---

[8]See Shiller (1981b) or Kleidon (1986b). An example may be helpful in this context. Suppose fundamentals are expected to be constant over time, i.e. $E[x_{t+i} \mid I_t] = x$ $\forall i > 0$. The ex-ante rational endogenous variable y_t then equals $y_t = \frac{1}{1-\alpha}x$ and is constant, as well. Notice that the corresponding variance $Var(y_t)$ is zero. Unexpectedly, in period $t + s > t$, $s > 0$ the actual fundamental stream lies above the expected, i.e. $x_{t+s} = x + \Delta$ with $\Delta > 0$ independent of x and returns to its normal value one period later. Then, the ex-post rational endogenous variable equals $y_t^* = \frac{1}{1-\alpha}x + \frac{1}{\alpha^{1+s}}\Delta$ during period t to $t + s$. This leads to a variance of y_t^* which lies above zero, establishing the required inequality.

[9]Orthogonality tests are formulated in Theorem 3 of LeRoy, Porter (1981).

Appropriate test statistics and their distributional properties which usually differ between the various types of variance bounds tests are derived for both original variance bounds tests and orthogonality tests in order to gain evidence to the significance of the violation of the null hypothesis[10].

The results of first variance bounds tests were striking: most of them rejected the null hypothesis of market efficiency. For the case of efficient capital markets, two characterizations were examined: The efficiency of the term structure of returns on bonds[11], based on different variance bounds and different timing of returns, were investigated by Shiller (1979, 1981b) and Singleton (1980). They find that the various inequalities, implied by the hypothesis of efficient capital markets, are violated. Concerning the efficiency of stock markets, Shiller (1981c) shows that by using annual price and dividend data from 1871–1979 Standard & Poor index, the null hypothesis of market efficiency formulated in equation (4.3) is dramatically violated: the variance of the ex–ante rational endogenous stochastic process reveals to be five times higher than the variance of the ex–post rational endogenous stochastic process. LeRoy, Porter (1981) obtained similar results using appropriately adjusted annual earnings as fundamentals. For the case of exchange rates, Huang (1981) tests for efficient exchange rates by using the monetary approach to balance of payments, presented in the previous chapter, and rejected the hypothesis of efficiency for the $\$/\pounds$, $\$/DM$ and $DM/\pounds$ exchange rates during the period 1973–1979[12].

Again, it should be mentioned that first variance bounds tests are derived under the joint assumption of (i) rational expectations, (ii) constant parameter α (which, for instance, implies a constant discount rate in the case of efficient capital markets) (iii) risk–neutrality and (iv) stationarity of the fundamental process. Notice that the first three assumptions imply the validity of equation (2.4), i.e. the assumption of efficient markets. The

[10]Concerning the precise nature of the test statistics, we refer to LeRoy, Porter (1981) or the literature cited above.

[11]Market efficiency with respect to returns on bonds implies that a long term yield should be representable by a weighted average of short term rates.

[12]See Diba (1987), however, for a critical discussion of variance bounds tests in the context of exchange rates, disputing mainly the approach applied by Huang (1981). The criticism concerns the semi–elasticity of interest rates used in these studies.

assumption of the stationarity of the fundamental process enables us to compute the sample variances on the basis of the sample data of y_t and x_t as an asymptotically unbiased estimator of the population variance that forms part of the null hypothesis formulated above. Precisely this joint hypothesis makes the interpretation of the violation of variance bounds difficult. Consequently, a subsequent discussion focuses on these joint assumptions.

Various critical statements of first variance bounds tests were mainly based on statistical problems[13]. Flavin (1983) and later Kleidon (1986a) state that the null hypothesis formulated in equation (4.3) is biased towards rejection in small samples. This is due to the fact that by using sample mean, instead of population mean, computation of the sample variance is always downward biased in comparison to the population variance if the sample size is small[14]. The variance of the sample mean will then capture some of the stochastic nature of the time series considered. Since the stochastic process $\{y_t^*, t \in \mathcal{T}\}$, as a moving average, is highly, serially correlated the variance of the sample mean of y_t^* exceeds the variance of the sample mean of y_t. establishing the bias of the difference $Var(y_t^*) - Var(y_t)$ against rejection. Hence, a violation of the inequality, stated in the null hypothesis above, may be misleading in small samples[15].

While both Flavin (1983) and Kleidon (1986a) hold to the assumption of stationarity of the fundamental process, Marsh, Merton (1986) criticize

---

[13]LeRoy, LaCivita (1981) emphasize that the violation of the variance bounds tests can also be caused by risk-averse individuals. Their paper states that the results of variance bounds can only be correctly interpreted if low risk-aversion or risk-neutrality prevails.

[14]Flavin derived a significant downward bias for sample sizes of less than 100 observations.

[15]In a later comment, Shiller (1986a), however, pointed out that, intuitively, the bias identified by Flavin (1983) is too small as to justify the high violation of the tested variance bounds, at least in the case of stock prices. In fact, Flavin (1983) examines excess volatility of short run return on bonds in comparison to their long run counterpart. Kleidon (1986a) calculates the bias of variance bounds tests in the framework of stock prices by using simulation techniques. He admits, however, that the downward bias is not sufficient to cause such high violation of the variance bounds, at least for stock prices.

precisely this premise of first variance bounds tests. When testing jointly for the validity of rational expectations, a constant discount rate, risk-neutrality of individuals and stationarity of the fundamental process[16], a rejection of the null hypothesis of first variance bounds tests, however, does not necessarily contradict the market efficiency hypothesis. In contrast, first variance bounds tests are highly sensitive with respect to the stationarity assumption of the fundamental process[17]. In the context of efficient capital markets Marsh, Merton (1986) consider a non-stationary fundamental process (i.e. a non-stationary dividend process) which grows at a constant rate, includes, however, additionally a component that reacts to changes in permanent earnings of a firm in comparison to the long-run growth rate of earnings[18]. A variance bound is obtained that contradicts the null hypothesis of first variance bounds tests in the sense that it states the converse inequality. Marsh, Merton (1986) conclude that the violation of the inequality of first variance bounds tests does not necessarily contradict the market efficiency hypothesis. Accepting the assumptions of rationality, constant discounting and risk-neutrality, a rejection of the null hypothesis can also be due to the (false) assumption of stationarity of the fundamental process. Therefore, a violation of the variance bounds

---

[16]To be more precise, this stationarity argument strikes mainly Shiller's (1981) analysis where the fundamental process is assumed to be stationary after detrending the time series by its average growth rate. LeRoy, Porter (1981), however, use a different trend correction such that the above argument does not seem to be applicable to their analysis.

[17]Kleidon (1986b), for instance, compares a plot containing time paths of actual price data and perfect foresight prices, as first described by Shiller (1981b) with a plot containing non-stationary price series and their perfect foresight counterpart, simulated by using a geometric Brownian motion for the fundamental process. Both plots look quite similar. Moreover, calculating the ex-post rational value based on fundamentals following a geometric random walk, and applying the variance bounds tests suggested by Shiller (1981b), the null hypothesis of prevailing market efficiency was rejected in 90 % of simulations conducted. Shiller (1988b) later admits that the choice of the simulation parameters was too extreme. A more appropriate choice in accordance with the actual movement of dividends will result in lower rejection rates.

[18]In the context of efficient capital markets no optimal dividend policy exists, as was pointed out by Miller, Modigliani (1961). Based on survey data, Marsh, Merton (1986) assume a dividend policy which tries to capture some stylized facts of dividend rules supposed to be applied by managers.

inequality does by no means imply the inefficiency of capital markets since "for any set of stock market price data, the hypothesis of market rationality can be rejected by some variance bounds test." (Marsh, Merton (1986))

These two central issues against first variance bounds tests (biasedness and non–stationary fundamentals) inspired the development of a *second generation of variance bounds test.* Mankiw et al. (1985), for instance, proposed such a new test by taking into account the shortcomings of first variance bounds tests[19]. They consider what they call a 'naive' forecast of the fundamentals denoted by $F[x_{t+i} \mid I_t]$ for $\forall i \geq 0$ conditional on information set I_t, and center the sequences of y_t and y_t^* around a corresponding naive forecast for the endogenous variable. Thus, define a naive forecast of the endogenous variable y_t^0 as the familiar present value relation consisting of naive forecasts of the fundamentals

$$y_t^0 = \sum_{i=0}^{\infty} \alpha^i F[x_{t+i} \mid I_t]$$

Notice that y_t^0 does not need to be efficient or rational in any sense. Now the following identity can be stated:

$$(y_t^* - y_t^0) = (y_t^* - y_t) + (y_t - y_t^0) \tag{4.4}$$

The first difference on the right side is the above derived error term u_t consisting of an infinite sum of rational prediction errors. Since u_t is uncorrelated with any element of I_t, the first difference on the right side is uncorrelated with the second difference on the right side, i.e. $E[(y_t^* - y_t)(y_t - y_t^0) \mid I_t] = 0$. Thus, taking squares on both sides of equation (4.4) and taking conditional expectations, we obtain

$$E[(y_t^* - y_t^0)^2 \mid I_t] = E[(y_t^* - y_t)^2 \mid I_t] + E[(y_t - y_t^0)^2 \mid I_t] \tag{4.5}$$

By the law of iterated expectation, equation (4.5) can be downdated establishing an expression which appears to consist of conditional variances, but instead of the sample mean, the naive and the ex–ante rational forecast

[19]For a good derivation of Mankiw et al. (1985) variance bound tests see also Diba (1990).

are used as reference levels. Thus, it can be interpreted as a composition of forecast errors defined previously.

$$(4.6) \qquad E[(y_t^* - y_t^0)^2 \mid I_{t-i}] = E[(y_t^* - y_t)^2 \mid I_{t-i}] + E[(y_t - y_t^0)^2 \mid I_{t-i}] \quad \forall i \geq 0$$

Notice that by arriving at equation (4.6) no homoskedasticity and no stationarity of $(y_t^* - y_t)$ and $(y_t - y_t^0)$ are required. Equation (4.6) implies two testable inequalities[20]

$$(4.7) \qquad E[(y_t^* - y_t^0)^2 \mid I_0] \geq E[(y_t^* - y_t)^2 \mid I_0]$$

$$(4.8) \qquad E[(y_t^* - y_t^0)^2 \mid I_0] \geq E[(y_t - y_t^0)^2 \mid I_0]$$

Equation (4.7) says that the ex–post rational stochastic process $\{y_t^*, t \in \mathcal{T}\}$ is better forecasted by the ex–ante rational stochastic process $\{y_t, t \in \mathcal{T}\}$ than by the naive forecast y_t^0. Equation (4.8) says that the ex–post rational stochastic process $\{y_t^*$ fluctuates more extensively around the naive forecast y_t^0 than does the ex–ante rational stochastic process $\{y_t, t \in \mathcal{T}\}$.

Mankiw et al. (1985) show that these inequalities avoid the downward biasedness of the first variance bounds tests, as criticized by Flavin (1983). In addition, the Mankiw et al. (1985) test can be applied if the endogenous and/or the exogenous stochastic process are non–stationary, circumventing the problem pointed out by Marsh, Merton (1986)[21]. Interestingly, their tests for efficiency of capital markets also reject the null hypothesis for all reasonable values of the discount factor which they assume to be exogenous, although the impact of violation is smaller in comparison to first variance bounds tests.

---

[20] Here, $i = t$ is assumed. For $i = t$ and $y_{-i} = 0$, $\forall i > 0$, the corresponding information set I_0 contains the initial conditions of all time series involved.

[21] There exist a variety of alternative variance bounds tests which consider non–stationary fundamental processes. See, for instance, Meese, Singleton (1983) in the context of foreign exchange markets or West (1986) in the context of US–stock market. More recently, LeRoy, Parke (1992) proposed several variance bounds of the ratio of the exogenous and endogenous variable, mitigating the non–stationarity of processes involved. See also Kalyvitis, Pittis (1994) for the foreign exchange market. An exposition of variance bounds tests under the special aspect of stationarity of the fundamentals provides West (1988a)'s table I.

Despite the apparent solution of the special problems of first variance bounds tests through the approach proposed by Mankiw et al. (1985), there still remains a difficulty, due to data deficits if variance bounds tests are thought of as implicitly testing for rational bubbles[22]. Trying to compute the variance of the ex-post rational stochastic process $\{y_t^*, t \in \mathcal{T}\}$ on the basis of sample data of the exogenous stochastic process, the problem of calculating an infinite sum arises. Since an infinite sequence of fundamentals is not available in observable data, the infinite sum has to be truncated at the point of the last observation. Thus, in order to be able to calculate the value of y_t^*, the following proxy variable $\bar{y}_t$ is generally applied

$$(4.9) \qquad \bar{y}_t = \sum_{i=0}^{T-t-1} \alpha^i x_{t+i} + \alpha^{T-t} g(y_T, I_T) \approx \sum_{i=0}^{\infty} \alpha^i x_{t+i} = y_t^*$$

where T refers to the last observation available and $g(y_T, I_T)$ denotes an arbitrary function depending on the last observation y_T of the endogenous stochastic process and the previous sequences of all variables in question contained in I_T[23]. Usually, the last observation y_T of the endogenous stochastic process is used such that equation (4.9) becomes

$$(4.10) \qquad \bar{y}_t = \sum_{i=0}^{T-t-1} \alpha^i x_{t+i} + \alpha^{T-t} y_T$$

Flood, Hodrick (1986), however, show that the proxy variable $\bar{y}_t$ itself contains a bubble. Consider the ex-post rational stochastic process $\{y_t^*, t \in \mathcal{T}\}$ as defined in (4.1) and decompose y_t^* in order to rewrite equation (4.10) as

$$(4.11) \qquad \bar{y}_t = y_t^* + \alpha^{T-t}(y_T - y_T^*) \quad \text{with} \quad y_T^* = \sum_{i=0}^{\infty} \alpha^i x_{T+i}$$

According to the general solution of equation (2.4) and its martingale part, the last term on the right side of equation (4.11) can be identified with the

[22] Recall that variance bounds tests are applied in order to test for market efficiency. Nothing has been said about rational bubbles yet.

[23] Shiller (1981b) propose to use the sample mean for $g(x_T, I_T)$ which was adapted by Marsh, Merton (1986). But, as Flavin (1983) pointed out, using the sample mean is a main source for the biasedness of the sample variances in small samples. Mankiw et al. (1985) instead use the actual value of the endogenous stochastic process at time T avoiding the problems connected with biasedness of sample mean and sample variances.

bubble solution such that

$$\alpha^{T-t}(y_T - y_T^*) = \frac{1}{\alpha^t} M_t$$

It can easily be shown that the difference $(y_T - y_T^*)$ equals the expected value of the bubble term at time T, i.e.

$$\begin{aligned} \alpha^{T-t}(y_T - y_T^*) &= \frac{1}{\alpha^t} M_t \\ (y_T - y_T^*) &= \frac{1}{\alpha^T} M_t = \frac{1}{\alpha^T} E[M_T \mid I_t] \end{aligned}$$

where the last rearrangement holds because of the martingale property of M_t. In other words, the proxy variable $\bar{y}_t$, constructed in order to test for the null hypothesis of market efficiency, implying the absence of rational bubbles in view of a limited data set, contains itself a bubble component. A rejection of the null hypothesis can by no means be interpreted as an evidence for the existence of rational bubbles or – alternatively – a non-rejection does not provide any evidence against the existence of rational bubbles since the observable counterpart (4.11) includes a bubble in any case. Consequently, the test becomes invalid in small samples[24].

Finally, a comprehensive overview of variance bounds tests developed so far and their corresponding results is provided by West (1988a), focusing particularily on non-stationarity requirements of the processes involved. No unique answer can be given concerning market efficiency. However,

---

[24]West (1986, 1988b), however, proposes a variance bounds test which avoids the problem pointed out by Flood, Hodrick (1986). He derives an inequality in terms of innovations based on the idea that if no rational bubbles exist, the variance of the innovations concerning the information set I_t is less than or equal to the variance of the innovations concerning a subset H_t of I_t, i.e. $H_t \subset I_t$,

$$H_0 : \qquad \mathrm{Var}(E[y_t \mid I_t] - E[y_t \mid I_{t-1}]) \leq \mathrm{Var}(E[y_t \mid H_t] - E[y_t \mid H_{t-1}])$$

Notice that only the left side can be influenced by rational bubbles such that a violation of H_0 indicates the existence of rational bubbles. In West (1988b), the test, applied to US-stock market, rejects the hypothesis of efficient capital markets whereas in West (1986) some evidence for efficient foreign exchange markets [DM/$] is given. For a more detailed description of the test procedure see West (1986). Moreover, as LeRoy (1989), footnote 14 pointed out, the test procedure suggested by West (1986, 1988b) is similar to a variance bounds test, proposed by LeRoy, Porter (1981).

evaluating the performance of the different variance bounds tests, West concludes that despite the possibility of a small sample bias[25], these distortions will not be sufficient to explain the partly great violation of variance bounds. Consequently, he recommends to investigate determinants of the endogenous process beyond simple fundamental components.

Nonetheless, variance bounds tests have provided many deeper insights, above all concerning the performance of capital markets – their primary field of application. These commonly accepted insights shall now be summarized as a final assessment of variance bounds tests, taking simultaneously into account their relevance in testing for rational bubbles.

- Variance bounds tests usually reject the null hypothesis of market efficiency and lead, therefore, to the conclusion that asset prices do not seem to reflect their fundamentals correctly. A component, in addition to the present value of future fundamentals, is likely to be present in the data.

- Despite the above mentioned caveat of small sample bias in variance bounds tests, this bias alone is not responsible for the partly dramatic violation of theoretically derived variance bounds.

- Variance bounds tests should, however, be primarily considered as tests for market efficiency. A violation of a corresponding variance bound can only be interpreted as a first hint for the existence of rational bubbles. It may, also, be explained by other factors like time-varying risk-premia or discount rates as well as by a general misspecification of the entire model considered.

In sum, prices, at least on capital markets, appear to be excessively more volatile than justified by the underlying fundamentals and the presence of rational bubbles provides one possible cause for this deviation. However, no clear conclusion concerning the existence of rational bubbles can be drawn from variance bounds tests.

---

[25]This small sample bias may be the reason why variance bounds tests, although initially very much in fashion, became less attractive and almost disappeared at the end of the 1980s.

4.1.2 Specification tests

Unlike variance bounds tests, specification tests focus more directly on the possible existence of rational bubbles. The central idea of specification tests developed by West (1987) is to consider two sets of consistent parameter estimates whereas only one set admits the existence of rational bubbles and then to conduct a Hausmann specification test in order to examine whether the estimates of both sets differ significantly or not. A rejection of the null hypothesis that there is no difference between both sets, may be interpreted as a hint for the existence of rational bubbles.

The first set of estimates is derived by estimating the parameter of the fundamental solution alone. Consistent estimates are obtained by simply applying OLS–estimation techniques. The corresponding forward solution is calculated by using proposition 2.3 and 2.4 respectively. This requires, however, a precise description of the fundamental process $\{x_t, t \in \mathcal{T}\}$ which can usually be obtained by conducting the familiar Box–Jenkins procedure of time series analysis[26]. Notice that by computing merely the fundamental solution, a possible existence of rational bubbles is excluded.

The second set of estimates is obtained by estimating the linear stochastic difference equation (2.4) directly, taking simultaneously into account the ARMA–representation already derived for the fundamental process. Equation (2.4) can be estimated by using instrumental variable techniques, as proposed by McCallum (1976). By using the common property of rational· expectations, equation (2.4) can then be rewritten as

$$y_t = \alpha y_{t+1} + x_t + \alpha \epsilon_{t+1} \tag{4.12}$$

where ϵ_{t+1} refers to the rational prediction error at time $t + 1$. In order to estimate equation (4.12) consistently, lagged values of the fundamental process or the endogenous process can be used as instruments for y_{t+1}. The ARMA–representation of the fundamental process can be estimated by standard least squares where the variance/covariance matrix is usually corrected for possible heteroscedasticity. Notice that this model specification admits the existence of rational bubbles.

[26]This procedure tries to represent x_t as $\Phi(L)x_t = \Theta(L)\epsilon_t$ where $\Phi(L)$ is a polynomial in L of degree p and $\Theta(L)$ is a polynomial in L of degree q.

If both model specifications describe reality adequately, i.e. if no rational bubbles exist, some restrictions concerning the estimated parameters of both model specifications must hold. These restrictions under the null hypothesis of no rational bubbles can be summarized by $R(\hat{\theta}) = 0$, where $\hat{\theta}$ denotes the whole set of estimates of both models. The null hypothesis can then be tested by the familiar Hausman test that proposes the following test statistic

$$W(\hat{\theta}) = R(\hat{\theta})' \left[\left(\frac{\partial R}{\partial \hat{\theta}} \right) \mathrm{V} \left(\frac{\partial R}{\partial \hat{\theta}} \right)' \right]^{-1} R(\hat{\theta}) \sim \chi^2(n+1)$$

where V denotes the asymptotic variance/covariance matrix. The test statistic $W(\hat{\theta})$ is asymptotically χ^2-distributed with $n + 1$ degrees of freedom where n is the number of restrictions to be tested.

As West (1987) pointed out, a specification test should only be applied if diagnostic tests supports the ARMA–representation of the fundamental process and the stochastic linear difference equation (2.4) since any difference between the two sets of estimates should be ascribed to sampling errors. In particular, non–stationarity of fundamentals requires a change in the test procedure. An appropriate test procedure in this case is also suggested by West (1987).

The empirical evidence of specification tests is mixed: West (1987) himself examines the US–stock market based on two different data sets from 1871 to 1981 and finds out that the null hypothesis of no rational bubbles is rejected by the data. Dezhbakhsh, Demirguc–Kunt (1990), however, do not reject the hypothesis of no rational bubbles for the same data set by using alternative test statistics, instead of Hausmans', as proposed by West (1987).

For the case of exchange rates Meese (1986) first examines the possible existence of rational bubbles on foreign exchange markets by using Wests' specification test. He finds evidence for the presence of rational bubbles within the $US\$/DM$ and the $US\$/£$ rates using monthly data over the period 1973–1982[27].

---

[27]In addition, Kearney, MacDonald (1990) test for the presence of rational bubbles in the Australian \$/US\$ rate during 1984–1986 and do not reject the null hypothesis of no rational bubbles.

For the case of price level rational bubbles Casella (1989) who independently developed the same specification test as West (1987) carries out a specification test for the presence of rational bubbles during the German hyperinflation (1920–1923) under the assumption of exogenous and endogenous money supply. While in the former case evidence in favour of the existence of rational bubbles is found, the null hypothesis of no rational bubbles can be rejected in the latter case[28].

However, evaluating the performance of specification tests, the following caveats have to be stressed:

- It seems that the results of specification tests depend on the set of instruments used to estimate equation (4.12) consistently. Different sets of instruments tend to change the outcome[29].

- As Flood, Hodrick, Kaplan (1986) pointed out, there exist a difference between estimation of parameter α based on a one–period return model (2.4) and based on a multi–period characterization of a return on assets. They find significant changes in the rate of return if a two–period model is used.

  This may be interpreted as evidence against the constant rate of return assumption underlying the model presented above. However, this criticism applies to all models described here and readopts the arguments already discussed in the last chapter.

- The model is sensitive concerning the description of the fundamental process, as Flood, Hodrick (1990) emphasize. They argue that dividends are better explained by being stationary in log–differences than in levels or pure differences. Additionally, a more complex ARIMA–process may perform better.

- Dezhbakhsh, Demirguc-Kunt (1990) also stress the low power of the Hausman test–statistic and apply a more powerful difference– and

[28]In addition, Engsted (1993b) tests for hyperinflationary rational bubbles in Israel, Brazil and Argentina for the second half of the 1980s, but does not find any evidence in either case.

[29]See Casella (1989).

reset–tests in order to test for a significant difference in the estimated parameters.

Despite the above mentioned problems, West's specification test can be considered as the first convincing attempt to investigate the existence of rational bubbles empirically. However, the general low power of indirect test procedures which we will discuss in greater detail at the end of this chapter is equivally valid here. Nonetheless, specification tests are applied in recent studies, but rather complementary than exclusive.

4.1.3 Integration and cointegration tests

Unlike variance bounds tests and specification tests, which often presume stationary or detrended variables, integration and cointegration tests concentrate more directly on the non–stationarity property of economic time series. Both tests for integration and tests for cointegration rely on the fact that bubble processes are non–stationary by construction. Due to this non–stationarity property of rational bubbles, restrictions concerning the degree of (co–)integration of the endogenous and fundamental process can be derived.

Before discussing the relevance of the integration and cointegration test in the context of rational bubbles, some general remarks concerning the concept of integration and cointegration, including estimation and testing techniques, should be pointed out[30].

Since many economic time series appear to be non–stationary and most of the estimation techniques are only applicable for stationary time series, the concept of integrated processes became important in econometric analysis.

Definition 4.1 A stochastic process is said to be *integrated* of order d, denoted by $I(d)$, if it has a stationary, invertible non–deterministic ARMA– representation after differencing d times, i.e. a process y_t is I(d) if $(1 - L)^d y_t$ is stationary with L denoting the first order difference operator.

[30] For an extended description of estimation and testing techniques see section 5.3.

It can be shown that most of economic time series appear to be I(1)-variables, i.e. stationary after differencing once[31]. Econometric analysis, which should be based on stationary data, becomes an analysis of differenced variables avoiding problems of non–stationary regressors if variables are expressed in levels. Unfortunately, by using merely variables in differences, information provided by variables expressed in levels, especially long run relationships between variables get lost.

This leads to the idea of cointegrated variables which tries to provide a connection between variables in levels and in differences. The concept of cointegration which was first introduced by Engle, Granger (1987), is stated here for the bivariate case of two random variables x_t and y_t.

Definition 4.2 The random variables x_t and y_t are said to be *cointegrated* of order d, b, denoted by $CI(d,b)$, if (i) both processes are $I(d)$ and (ii) if there exists a parameter $\vartheta \neq 0$ such that

$$z_t = y_t - \vartheta x_t \tag{4.13}$$

is integrated of order $d - b$, $b > 0$. The vector $(1, \vartheta)$ is called the cointegrating vector.

The economically most interesting case is $d = 1,\ b = 1$, i.e. both variables are non–stationary in levels and stationary in first differences, but their linear combination z_t is stationary in levels. Then, equation (4.13) is often interpreted as an equilibrium error where ϑ corresponds to the long–run equilibrium relationship between y and x. Cointegration, implying the stationarity of the equilibrium error z_t, means that the deviation from the long run relationship between y and x rarely drift far apart from zero while the deviation of z_t from its equilibrium could possibly be rather large when x and y are not cointegrated.

Focusing on $CI(1,1)$–processes, an interesting proposition has been proven by Engle, Granger (1987) and extended by Engle, Yoo (1987) called Granger Representation Theorem[32]:

[31] See Nelson, Plosser (1982).

[32] Here, we only state the bivariate case. The original theorem is formulated for the multivariate case.

Proposition 4.1 If two variables y_t and x_t are $CI(1,1)$, i.e. $z_t = y_t - \vartheta x_t$ is stationary, then there exists an *error correction model (ECM)* of the following form:

$$\Delta y_t = \gamma_0 + \gamma_1 z_{t-1} + \sum \gamma_{1i} \Delta x_{t-i} + \sum \gamma_{2i} \Delta y_{t-i} + \epsilon_t \quad (4.14)$$

where Δ denotes the first order difference and ϵ_t a sequence of iid random variables with zero mean and variance σ_ϵ^2.

Proof of proposition 4.1 See Engle, Granger (1987). ||

Note that the ECM–representation (4.14) is formulated in levels and differences, including, however, only stationary regressors if the equilibrium error z_t is used explicitly and cointegration prevails. The ECM–representation captures the idea that a change in the dependent variable y_t is influenced not only by the change of x_t and y_t of previous periods, but also by the extent of disequilibrium in the preceding period characterized by the equilibrium error z_t. γ_1 represents the reaction coefficient in face of disequilibrium. At least for the bivariate case, we expect γ_1 to be smaller than zero. Otherwise, the behaviour of the system would not be stable in the sense that deviations from the equilibrium in one period lead to a correction of the error in the following period.

Now we return to the description of integration and cointegration tests in the context of rational bubble: Based on their general critical assessment of the theoretical possibility of rational bubbles in a rational expectation framework[33], Diba, Grossman (1988a), Hamilton, Whiteman (1985) and Hamilton (1986) propose to leave off the area of parametric test procedures and to test only for the degree of *integration* of the endogenous stochastic process and the fundamental process. The rationale of integration tests can be shown by considering the general solution of equation (2.4), i.e. $y_t = f_t + b_t$ where f_t denotes the fundamental solution and b_t the bubble solution. According to equation (2.37) of proposition 2.12, the bubble process can be characterized by

$$E[b_{t+1} \mid I_t] = \frac{1}{\alpha} b_t \quad (4.15)$$

[33] We turn to this issue more extensively in a later section.

Notice that for $0 < \alpha < 1$, which is the economically relevant case, both the expectation and the variance of b_t increases without end if t approaches infinity establishing the non–stationarity property of the bubble process b_t[34]. Hence, if rational bubbles, according to equation (4.15) non–stationary by construction, in fact exist, then the endogenous stochastic process $\{y_t, t \in \mathcal{T}\}$ has to be integrated of higher order than the corresponding fundamental process $\{x_t, t \in \mathcal{T}\}$. Thus, if the endogenous stochastic process $\{y_t, t \in \mathcal{T}\}$ is integrated of order d, denoted by $I(d)$ and the fundamental process f_t is integrated of order b, denoted by $I(b)$, then the null hypothesis of no rational bubbles would state that the endogenous stochastic process should be integrated of the same degree as the fundamental process, i.e.

$$H_0: \quad d = b \text{ if } y_t \text{ is } I(d) \text{ and } x_t \text{ is } I(b)$$

Notice that if the null hypothesis cannot be rejected, rational bubbles cannot exist, due to their non–stationarity property. Hence, the purpose of integration tests is to confirm rather the absence of rational bubbles than their presence. However, rejecting the null hypothesis does not necessarily mean that rational bubbles exist. As Diba, Grosman (1988a), Hamilton, Whiteman (1985) and Hamilton (1986) emphasize, any non–stationary variable unobserved by the econometrician can cause the deviation of the endogenous stochastic process from its fundamentals[35].

Meese (1986) and Diba, Grossman (1988a), among others, suggest to test for *cointegration* of the endogenous stochastic process and the fundamental process, in addition to integration tests. Unlike integration tests, tests for cointegration postulate a precise description of the underlying economic model, i.e. a structural relation between endogenous and fundamental process. The precise nature of the fundamental process, however, does not need to be specified.

The idea of cointegration tests can best be shown by decomposing the well–

[34]This non–stationarity can also be shown by using simulation techniques. Diba, Grossman (1988a), for instance, conduct a simulation based on equation (4.15) for reasonable values for α. By computing the autocorrelations up to lag 10, they confirm the non–stationarity by using standard techniques of time series analysis.

[35]For a more detailed discussion of this issue see subsection 4.2.3.

known forward solution which normally forms the basis of cointegration tests.

$$\begin{aligned}
f_t &= x_t + \alpha E[x_{t+1} \mid I_t] + \alpha^2 E[x_{t+2} \mid I_t] + \dots \\
&= x_t + \alpha E[\Delta x_{t+1} \mid I_t] + \alpha x_t + \alpha^2 E[\Delta x_{t+2} \mid I_t] \\
&\quad + \alpha^2 E[\Delta x_{t+1} \mid I_t] + \alpha^2 x_t \dots \\
&= x_t + \alpha x_t + \alpha^2 x_t + \dots \\
&\quad + \alpha(E[\Delta x_{t+1} \mid I_t] + \alpha E[\Delta x_{t+1} \mid I_t] + \alpha^2 E[\Delta x_{t+1} \mid I_t] + \dots) \\
&\quad + \alpha^2(E[\Delta x_{t+2} \mid I_t] + \alpha E[\Delta x_{t+2} \mid I_t] + \alpha^2 E[\Delta x_{t+2} \mid I_t] + \dots) + \dots \\
&= \frac{1}{1-\alpha} x_t + \frac{\alpha}{1-\alpha} E[\Delta x_{t+1} \mid I_t] + \frac{\alpha^2}{1-\alpha} E[\Delta x_{t+2} \mid I_t] + \dots \\
&= \frac{1}{1-\alpha} x_t + \frac{1}{1-\alpha} \sum_{i=1}^{\infty} \alpha^i E[\Delta x_{t+i} \mid I_t]
\end{aligned}$$

Simple rearrangement yields the desired cointegration relation.

$$y_t - \frac{1}{1-\alpha} x_t = \frac{1}{1-\alpha} \sum_{i=1}^{\infty} \alpha^i E[\Delta x_{t+i} \mid I_t] \tag{4.16}$$

Notice that in most of the empirical analyses it turns out that both the endogenous stochastic process and the fundamental process are integrated of order 1. Consequently, the first difference of the fundamental process, appearing on the right side of equation (4.16), is stationary. Thus, we can infer that in the absence of rational bubbles - recall that only the fundamental solution is considered above – the linear combination of the endogenous stochastic process and the fundamental process, appearing on the left side of equation (4.16) should be stationary, too. If both the endogenous and the fundamental process are $I(1)$, and their linear combination is stationary, endogenous and fundamental processes are said to be cointegrated of order $CI(1,1)$, providing evidence against the presence of rational bubbles in the data. Therefore, cointegration tests state under the null hypothesis of no rational bubbles that endogenous and fundamental process are cointegrated of order $CI(1,1)$, i.e.

$$\begin{aligned}
H_0: \quad & y_t \text{ and } x_t \text{ are } CI(1,1) \text{ or} \\
& z_t = y_t - \vartheta x_t \sim I(0) \\
& \text{if } y_t \text{ and } x_t \text{ are } I(1)
\end{aligned}$$

where z_t denotes the corresponding residuals when regressing y on x. Moreover, in the corresponding regression equation trend and seasonal components are usually included[36].

The relation $y_t - \frac{1}{1-\alpha}x_t$ in equation (4.16) is usually called the 'spread'[37], denoted by S_t. Thus, the absence of rational bubbles can be verified by showing the stationarity of the spread. According to equation (4.16), the spread can be interpreted as a weighted average of rationally expected future changes in the fundamental process $\{x_t, t \in \mathcal{T}\}$. Equivalently, it can be shown that S_t is proportional to the expected change in the endogenous variable y_t for the next period, i.e. to $E[\Delta y_{t+1} \mid I_t]$. Hence, the spread can be viewed as an optimal forecast of expected changes in the endogenous variable.

If the sequence of expected changes in the fundamentals is stationary, then according to equation (4.16), the spread is stationary under the null hypothesis of no rational bubbles, and so does the sequence of future changes in the endogenous variable. Hence, if the endogenous variable y and the fundamentals x are cointegrated of order $(1,1)$ as indicated by a stationary spread in equation (4.16), then future changes in the endogenous variable are unpredictable or – equivalently – expected to be zero on average, which is another interpretation of a stationary spread in the absence of rational bubbles[38].

Moreover, while cointegration is usually interpreted as a long run equilibrium relationship between the variables considered, the concept of cointegration gets another interpretation within the context of present value models: the spread, theoretically derived above as a relation between cointegrated variables, equals the infinite sum of discounted expected future changes in the fundamentals or – equivalently – the expected change in

[36]Notice that the linear combination of endogenous and fundamental process can theoretically be represented by equation (4.16) in the context of present value models. Thus, for the parameters of the null hypothesis an economically reasonable restriction on the cointegration vector may be imposed, i.e. $(1, \frac{1}{1-\alpha})$ which provides a sensibility check on economic grounds.

[37]See Campbell, Shiller (1987).

[38]Notice that this establishes the property of the unpredictability of the rate of return on capital markets if log stock prices and log fundamentals are used.

the endogenous variable in the next period, and can, therefore, be viewed as an optimal forecast of both. Hence, the cointegration relation in form of the spread implies an optimal forecast property rather than a long-run equilibrium[39].

Pure tests for integration and cointegration are frequently conducted in literature, but no common conclusion can be drawn from the data. Stock prices and dividend processes using US-data seem to be integrated of order 1, but no unique answer can be given concerning cointegration. Diba, Grossman (1988a) use integration and cointegration techniques for testing for rational bubbles on US-stock market. While integration tests found no evidence in favour of the existence of rational bubbles, results of cointegration tests remain ambigous[40]. The evidence for exchange rates is also

[39]See Campbell, Shiller (1988). Based on the spread derived above, Campbell, Shiller (1987) suggest another procedure for testing rational bubbles by using a VAR-approach. Their primary intension is to examine the validity of present value models and to formulate – without assuming endogeneity and exogeneity a priori – a VAR-model in terms of changes in the fundamentals Δx_t and the spread S_t which can be viewed as a counterpart of an error-correction-model in x and y

$$\begin{bmatrix} \Delta x_t \\ S_t \end{bmatrix} = \begin{bmatrix} a(L) & b(L) \\ c(L) & d(L) \end{bmatrix} \begin{bmatrix} \Delta x_{t-1} \\ S_{t-1} \end{bmatrix} + \begin{bmatrix} u_{1t} \\ u_{2t} \end{bmatrix}$$

u_{1t}, u_{2t} are uncorrelated error terms and $a(L)$, $b(L)$, $c(L)$, $d(L)$ denote p-lag polynomials with lag operator L with lag length p to be determined. A present value model imposes several cross equation restrictions on the VAR-representation which can be tested by using a Wald-test statistic, for instance. Furthermore, Campbell, Shiller (1987) construct a theoretical spread by using the parameters estimated in the VAR-model and compare this theoretical spread with the actual spread computed by using actual data. They conclude that the null hypothesis of the validity of the present value model (i.e. market efficiency) cannot be rejected for the term structure, but is very unlikely to hold for the stock market. The VAR-test procedure introduced by Campbell, Shiller (1987) provides a natural extension for testing for the validity of present value models and implicitly for rational bubbles in the case of non-stationary endogenous and fundamental processes and is subsequently often applied in various economic fields. See Engsted (1993a), for instance.

[40]Interestingly, Diba, Grossman (1988a) conduct tests for cointegration by estimating the spread by OLS and use the critical values of Engle, Granger (1987) to test for cointegration, i.e. for stationarity of the residuals. These critical values, however, are derived by considering a constant in the cointegration relation which is not included in the spread. While both Campbell, Shiller (1988) and Diba, Grossman (1988) use annual

mixed: most exchange rates appear $I(1)$ but cointegration differs depending on the underlying fundamental process[41].

The following critical remarks concerning the performance of integration and cointegration tests may be pointed out:

- On the one hand, integration tests have the advantage that they do not explicitly presume the validity of a completely specificied economic model (like variance bounds tests which are based on the present value model). Insofar, the power of integration tests is less affected by a possible misspecification of an underlying economic model. On the other hand, for conducting integration tests the variables which can be viewed as fundamentals have to be determined, and these fundamental variables often rely on a certain economic model. This implies that, although integration tests do not require a specific economic model, integration tests always contain a hypothesis concerning the underlying fundamentals and are, therefore, sensible concerning misspecified fundamentals, as well.

- In particular, integration tests seem to have low power which is due to the fact that most of economic time series appear to be integrated of order 1[42]. Consequently, integration tests may mainly serve as a first check of the bubble hypothesis which has to be supplemented by other test procedures.

- The basis of integration and cointegration tests in favour of or against rational bubbles is the non–stationarity of the bubble process as indicated by equation (4.16). However, various studies show that bubble

data, Dwyer, Hafer (1990) apply data with shorter periodicity and find no cointegration between stock prices and fundamentals.

[41]The mixed evidence may also be due to the fact that different exchange rates were investigated in different periods. Kearney, MacDonald (1990) examine the Australian $\$/US\$$ rate between 1984–1986 within various exchange rate models and find that both endogenous and fundamental process are integrated of order 1 but not cointegrated. Pittis (1993) also conclude that exchange rates and differently defined fundamentals are $I(1)$. However, again no cointegration is found. See also Smith (1989).

[42]See Gaab (1990) for this criticism.

processes, although non–stationary by construction, do not necessarily imply an actual non–stationary sequence.

At first, Meese (1986) has carried out a simulation of a bubble process specified as a bursting bubble. He shows that by using standard techniques of time series analysis (calculating autocorrelation functions), the time path corresponding to a bursting bubble may be represented by a (stationary) autoregressive process such that the assumption of non–stationarity of the bubble process cannot hold under certain bubble specifications. Subsequently, Evans (1991) also uses a periodically bursting bubble and simulated 200 bubble processes, generated by the proposed bubble process, each of sample size 100. By applying standard techniques of time series analysis, he finds out that rational bubbles can be represented by a linear autoregressive, i.e. a stationary process, when bubbles are characterized appropriately. Consequently, the results of standard tests of integration and cointegration, that often do not lead to a rejection of the null hypothesis of rational bubbles, cannot be viewed as evidence in favour of rational bubbles. He concludes that the "presence of rational bubbles in actual stock prices thus remains an open question" (Evans (1991))

- Charemza, Deadman (1995) recently emphasize that traditional integration and cointegration tests in favour of or against rational bubbles assume a false null hypothesis. While common integration and cointegration tests rely on the null hypothesis of a random walk tested against stationary alternatives, a correct null hypothesis in the context of testing for rational bubbles should include the bubble process itself.

  Charemza, Deadman (1995) construct the following bubble process which they call a stochastic explosive root (STER–) process in its most general form

$$(4.17) \qquad \begin{aligned} B_t &= \exp(\theta_t) B_{t-1} \exp(u_t) \\ &\theta_t \sim IIN(\ln(1+r) - \frac{\sigma_\theta^2}{2}, \sigma_\theta^2) \\ &u_t \sim IIN(-\frac{\sigma_u^2}{2}, \sigma_u^2) \end{aligned}$$

$$E[\theta_t u_t] = 0$$

According to Charemza, Deadman (1995), a correct null hypothesis for testing for the existence of rational bubbles would postulate a process to follow (4.17) and not a random walk process as it is done in standard unit root tests. As a consequence, the critical values, derived by using the above bubble process as null hypothesis, are calculated by Monte–Carlo methods. They vary within a wide range and differ dramatically from those applied in standard unit root tests.

Subsection 4.1.5 provides a more extended description of the Charemza, Deadman (1995) analysis and discusses their validity. The results of our assessment can be summarized as follows:

- The problem of highly varying critical values can be mitigated by simply taking logs rather than examining time series in levels.
- The extreme deviation of critical values associated with the Charemza, Deadman (1995) bubble refers at least partly to economically unreasonable parameter combinations.
- Interestingly, the Charemza, Deadman (1995) bubble process, instead of providing a rather general STER–process, reveals to be a subcase of the GSTM–process previously derived. Therefore, the analysis of chapter 2 applies accordingly.

These results partly invalidate the Charemza, Deadman (1995) argument against integration or cointegration tests as tests for the existence of rational bubbles.

To date tests for integration and cointegration appear to be contained within any battery of tests for rational bubbles, but rather conducted complementarily than exclusively. They serve as a first check of the null hypothesis of no rational bubbles; a rejection, however, provides rather a necessary condition for the existence of rational bubbles than a sufficient condition.

4.1.4 Final assessment of indirect tests

Several indirect test procedures were presented above. The first indirect tests, i.e. variance bounds tests, can be viewed as being more concerned with the evaluation of the present value model than with trying to gain evidence for rational bubbles at all. Indirect test procedures, like specification tests and integration and cointegration tests, which focus more directly on rational bubbles, were discussed thereafter. All test procedures have been criticized above on grounds which in part are special for these test procedures. They have in common that they normally test for a joint hypothesis consisting of the already described assumption structure, i.e. rational expectations, constant parameter α, risk–neutral individuals[43]. These presumptions establish equation (2.4) as an adequate description of the world. Any failure of these presumptions would lead to a misspecified model. A possible rejection of the null hypothesis can be due to any of these presumptions as well as may indicate the existence of rational bubbles.

Besides this general caveat of possible misspecification, which principally meets any econometric model, the power of indirect tests in face of small samples proves to be unsatisfactory. In order to evaluate the performance of indirect test procedures, Mattey, Meese (1986) conduct a Monte–Carlo study by simulating six different model specifications, i.e. a model with time–varying discount rates, a dividend smoothing model, a model including a bursting bubble, a model capturing the idea of structural changes, and a present value model with stationary and non-stationary fundamentals. Based on these generated time series, each of 100 observations, different test statistics are calculated, among others the Mankiw et al. (1985) and West (1986, 1988b) variance bounds, West's (1987) specification test statistics and a battary of integration and cointegration tests in order to derive the distribution of the test statistic under varying model specifications. Each simulation was based on 500 replications.

The simulated distribution was compared with the theoretically expected distribution. It turned out that most of the test statistics were not distributed as theoretically warranted. Moreover, these small sample biases

[43] See the analysis in chapter 3.

reveal to be so large that a discrimination between the different model specifications becomes almost impossible. Since one of the model specifications contained a bursting bubble specification, it must be doubted that any indirect test procedure is likely to recover rational bubbles in the data and is simultaneously able to discriminate between different model specifications under question. This establishes a rather pessimistic view of indirect test procedures in the context of rational bubbles.

4.1.5 A digression: Charemza, Deadman (1995) analysis

As already mentioned earlier, Charemza, Deadman (1995)'s analysis of rational bubbles is based on the following bubble process

$$
\begin{aligned}
b_t &= \exp(\theta_t) b_{t-1} \exp(u_t) \\
&\theta_t \sim IIN(\ln(1+r) - \frac{\sigma_\theta^2}{2}, \sigma_\theta^2) \\
&u_t \sim IIN(-\frac{\sigma_u^2}{2}, \sigma_u^2) \\
&E[\theta_t u_t] = 0
\end{aligned}
\tag{4.18}
$$

Notice that b_t satisfies the required submartingale property of rational bubbles and a non–negativity condition such that (4.18) indeed can be viewed as a possible bubble process.

Now, according to Charemza, Deadman (1995), a correct null hypothesis for testing for the existence of rational bubbles in the context of integration and cointegration tests would consider the process (4.18) and not a random walk process as it is usually done in standard integration and cointegration tests.

Charemza, Deadman (1995) run a Monte–Carlo simulation of (4.18) of sample size 100 and regress the resulting time series by using OLS against[44]

$$b_t = \hat{\mu} + \hat{\beta}t + \hat{\rho}b_{t-1} + v_t \tag{4.19}$$

where $\hat{\mu}$ denotes the estimated constant, $\hat{\beta}t$ the estimated trend component and v_t the resulting residuals. Different test–statistics were calculated, among others the Fuller (1976) $t_{\hat{\rho}}$–statistic and the Dickey, Fuller (1981) Φ_3–statistic which test for the joint null hypothesis of $\beta = 0$ and $\rho = 1$[45]. The simulation of the bubble process b_t was repeated 25.000 times in order to get the distribution of the test–statistics involved, in particular $t_{\hat{\rho}}$. The Monte–Carlo study was conducted for the following set of parameters:

$$r = 0.03 \quad \sigma_u = 0.02 \quad \sigma_\theta = \{0.1; 0.15; 0.2; 0.25; 0.3; 0.4; 0.5\}$$

The resulting 5% critical values of the distribution of the $t_{\hat{\rho}}$ vary between -3.142 and -16.652. Charemza, Deadman (1995) conclude that traditional integration and cointegration tests are invalid in the context of rational bubbles since for statistical inference purposes the true null hypothesis should be a bubble process and, by using the Charemza, Deadman (1995) bubble process (4.18) as a null hypothesis, the critical values vary dramatically.

We repeat and extend the Monte–Carlo study initialized by Charemza, Deadman (1995) in the following way:

- A broader parameter set is used in order to get a deeper insight into the behaviour of the bubble process (4.18) and the power of integration tests. Holding σ_u constant at 0.02 the other parameters vary as follows:

$$\begin{aligned} r &= \{0.01; 0.02; 0.03; 0.04; 0.05; 0.06; 0.07; 0.08; 0.09; 0.10\} \\ \sigma_\theta &= \{0.01; 0.02; 0.03; 0.04; 0.05; 0.06\} \end{aligned}$$

[44]Obviously, Charemza, Deadman (1995) forget the trend term in their paper or report wrong critical values for the random walk hypothesis which clearly refer to the alternative (4.19). Another hint for the fact that they use (4.19), instead of a trendless regression, is that they apply the test–statistic Φ_3 of Dickey, Fuller (1981) which is based on (4.19).

[45]Notice that the Φ_3–statistic has a degenerated distribution tabulated in table V of Dickey, Fuller (1981), not in table VI, as stated by Charemza, Deadman (1995).

- Besides the level of the simulated series, their log-counterpart is tested for the presence of a unit root including a constant and a trend component.

Table 4.1: Results of Monte–Carlo simulation of Charemza, Deadman (1995) bubble

r	σ_θ					
	0.1	0.2	0.3	0.4	0.5	0.6
0.01	−3.515 (−3.443)	−4.587 (−3.471)	−7.759 (−3.442)	−12.301 (−3.474)	−20.461 (−3.445)	−30.368 (−3.450)
0.02	−3.327 (−3.454)	−4.803 (−3.449)	−6.594 (−3.431)	−11.566 (−3.448)	−18.589 (−3.426)	−29.646 (−3.435)
0.03	−3.156 (−3.471)	−3.810 (−3.442)	−5.661 (−3.486)	−9.979 (−3.440)	−17.481 (−3.439)	−27.596 (−3.460)
0.04	−2.828 (−3.449)	−3.602 (−3.461)	−4.838 (−3.461)	−9.046 (−3.457)	−16.147 (−3.448)	−26.102 (−3.458)
0.05	−2.390 (−3.462)	−3.504 (−3.452)	−4.430 (−3.440)	−7.807 (−3.470)	−14.732 (−3.460)	−23.900 (−3.485)
0.06	−1.901 (−3.453)	−3.360 (−3.446)	−4.141 (−3.439)	−6.948 (−3.451)	−13.113 (−3.469)	−22.504 (−3.437)
0.07	−1.363 (−3.452)	−3.217 (−3.447)	−3.996 (−3.463)	−5.972 (−3.447)	−11.729 (−3.447)	−21.608 (−3.461)
0.08	−0.786 (−3.459)	−3.056 (−3.455)	−3.865 (−3.467)	−5.286 (−3.445)	−10.480 (−3.469)	−18.815 (−3.455)
0.09	−0.229 (−3.433)	−2.891 (−3.456)	−3.747 (−3.455)	−4.883 (−3.421)	−9.613 (−3.471)	−18.799 (−3.450)
0.10	0.424 (−3.451)	−2.662 (−3.452)	−3.696 (−3.471)	−4.585 (−3.460)	−8.265 (−3.448)	−16.397 (−3.448)

Remarks: 5% fractiles of simulated Charemza, Deadman (1995) bubble, 25000 times, 5% fractiles of log-Charemza, Deadman (1995) bubble in parenthesis.

The resulting 5% percentiles of $t_{\hat{\rho}}$ are tabulated in table 4.1. The outcome of testing for the degree of integration, by using logs rather than levels, are reported in parenthesis. Notice that the Charemza, Deadman (1995) results for $r = 0.03$ are replicated.

Before discussing the results of the Monte–Carlo simulation in greater detail, we examine what can be expected from a theoretical point of view.

- First, by adequately combining the different random variables of the Charemza, Deadman (1995) bubble (4.18) and by separating deterministic and random components, we obtain because of independence of θ_t and u_t

$$b_t = b_{t-1} \exp(\mu + \eta_t) \tag{4.20}$$

$$\mu = \ln(1+r) - \frac{\sigma^2}{2}$$
$$\sigma^2 = \sigma_u^2 + \sigma_\theta^2$$
$$\eta_t \sim N(0, \sigma^2)$$

From equation (4.20) it is apparent that the Charemza, Deadman

Figure 4.1: Simulated Charemza, Deadman (1995) bubble

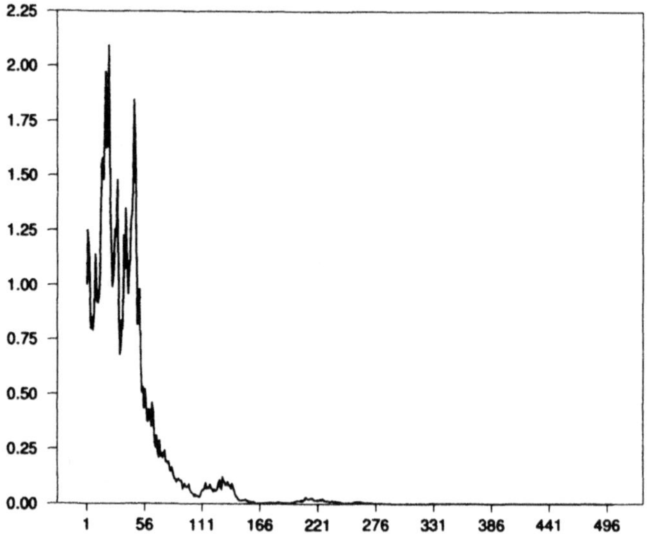

(1995) bubble is already contained in the set of bubbles derived in chapter 2[46]. The parameter μ stands for the expected or average growth rate of the bubble. It becomes obvious that, although theoretically possible, the Charemza, Deadman (1995) bubble shows an economically controversial behaviour if we allow for $\mu < 0$ because the bubble becomes arbitrarily small if time approaches infinity[47]. Hence, this type of bubble, although always present, has no decisive

[46] According to Charemza, Deadman (1995), the bubble processes defined by equation (4.18) represents a quite general stochastic explosive root (STER–) process which encloses other processes rather common in statistics (for instance, a geometric random walk). However, the converse result is established here: the STER–process is a subcase of the general GSTM–process derived in chapter 2.

[47] However, it never reaches zero by construction.

impact on the related endogenous process in the long run. This can be supported by figure 4.1 which shows a simulated time path of Charemza, Deadman (1995) bubble for $\mu = -1.0$ and $\sigma^2 = 0.04$. More exactly, we can state the following proposition:

Proposition 4.2 If the parameter μ of bubble specification (4.20) is smaller than zero, then the bubble process b_t becomes arbitrarily small.

Proof of proposition 4.2 We are looking for the probability that the bubble process b_t becomes smaller than an arbitrary $\epsilon > 0$ at $t > 0$, i.e.

$$\Pr\left[b_t < \epsilon\right] \quad \text{for } t \to \infty$$

From equation (4.20) we get by backward induction

$$b_t = b_0 \exp\left(\mu t + \sum_{i=1}^{t} \eta_i\right) \qquad \text{for } b_0 > 0$$

After substitution and dividing through $b_0 > 0$, we obtain

$$\Pr\left[\exp(\mu t + \sum_{i=1}^{t} \eta_i) < \frac{\epsilon}{b_0}\right]$$

This is the probability that the bubble shrinks at $\frac{\epsilon}{b_0}$ % of its initial value b_0. Simple rearrangements yield

$$\Pr\left[\mu t + \sum_{i=1}^{t} \eta_i < \ln \epsilon - \ln b_0\right]$$

$$\Pr\left[\mu < \frac{\ln \epsilon - \ln b_0}{t} - \frac{\sum_{i=1}^{t} \eta_i}{t}\right]$$

The first term on the right side approaches zero if time goes to infinity, i.e. plim $\frac{\ln \epsilon - \ln b_0}{t} \to 0$. Because of the stationarity of η_t, the second term on the right side approaches zero if time approaches infinity, i.e.plim $\frac{\sum_{i=1}^{t} \eta_i}{t} \to 0$. Thus, we obtain: the probability that the Charemza, Deadman (1995) bubble becomes arbitrarily small equals the probability that $\mu < 0$ which is 1 if μ is actually smaller than zero and zero if $\mu > 0$

$$\Pr\left[\mu < 0\right] = \begin{cases} 1 & \text{for } \mu < 0 \\ 0 & \text{for } \mu > 0 \end{cases}$$

Hence, we have shown that the probability that the Charemza, Deadman (1995) bubble becomes arbitrarily small is 1 if $\mu < 0$.

||

Now, it is questionable whether this shrinking behaviour of the Charemza, Deadman (1995) bubble is economically reasonable or whether an economically inspired restriction $\mu > 0$ should be introduced. We turn to this issue later on when discussing the simulation results.

- Second, taking logs of equation (4.20) yields a random walk with drift, i.e.

  (4.21) $$\ln b_t = \mu + \ln b_{t-1} + \eta_t$$

  where $\ln b_t$ denotes the log of Charemza, Deadman (1995) bubble. If the true data generating process is a random walk with drift and we test against a trend stationary alternative (4.19), the t–statistic $t_{\hat{\rho}}$ has a degenerated distribution and the corresponding critical values are tabulated in Fuller (1976). Note the distribution of $t_{\hat{\rho}}$ is independent of the parameter μ[48].

Simulation results reported in table 4.1 can be summarized as follows:

- Accepting the economically reasonable restriction $\mu > 0$, it becomes apparent that the extreme deviations from Fuller's (1976) critical value which is -3.45 for sample size 100, fall into the region where parameter combinations of r and σ_θ precisely violate this restriction. If we additionally impose another economically reasonable restriction $r < 0.1$, for instance, since average real returns higher than 10% are rarely observed, the variation of the critical values decrease considerably but still remains too large. The critical values lie within the

[48] If we, however, test against an autoregressive alternative solely containing a constant, the t—statistic $t_{\hat{\rho}}$ is distributed standard–normally in the limit. If the sample size is small, the distribution differs from the normal and its critical values vary with the parameter μ. The corresponding critical values for small sample sizes are tabulated in Hylleberg, Mizon (1989, 1995). For a more detailed description of unit root tests see section 5.3. See also appendix B for an exposition of the distributional properties of the various t–statistics in testing for the degree of integration.

interval $[-4.883; 0.424]$ which may be seen as an undecisible region. Consequently, integration tests based on these critical values have very low power.

- Testing for unit roots in logs rather than in levels proves to be more sucessful. Taking logs stabilizes the critical values around the theoretically warranted value of -3.45 (for T=100) since the true DGP is actually a the random walk with drift. The corresponding t-statistic is independent of the drift parameter as theoretically expected.

In sum, we recommend to test for a unit root in both levels and logs if it is suspected that the bubble process can be described by a Charemza, Deadman (1995) bubble process. This procedure will mitigate the shortcomings of integration and cointegration tests addressed by Charemza, Deadman (1995)[49].

4.2 Direct tests

Unlike indirect test procedures, the idea of direct tests is to consider explicitly a specific bubble process[50] and to test directly for the significance of the bubble solution. Insignificant parameters of the bubble solution would then only imply the absence of that specific bubble process whereas significant parameters would lead to the conclusion that the inclusion of that

---

[49]Besides the analysis of the bubble process (4.18), Charemza, Deadman (1992) investigate the characteristics of the critical values in the case of intrinsic bubbles, as introduced by Froot, Obstfeld (1991b). Again, they emphasize that, instead of using a random walk as null hypothesis, only the use of an appropriately defined intrinsic bubble formulation as null hypothesis is correct. By replicating and extending the analysis in the case of intrinsic bubbles, following the same lines as above, we again establish the result that taking logs will mitigate the problem of varying critical values. This outcome is reasonable since intrinsic bubbles are shown to be derived from our general GSTM-process to which the Charemza, Deadman (1995) bubble also belongs. Hence, the recommendation of taking logs applies here again.

[50]Of course, the choice of a specific bubble process among the infinite set of possible stochastic processes which fulfill $b_t = \frac{1}{\alpha} E[b_{t+1} \mid I_t]$ has to be justified by economic reasons. Otherwise, choosing a specific bubble process appears to be too arbitrary.

specific bubble process is likely to explain a greater part of endogenous variable's movement than simple present value models.

Due to the above mentioned difficulty in directly testing for rational bubbles (i.e. the multiplicity of the bubble solution) and the impossiblity (at least until now) of direct testing for wide classes of bubble processes (a bursting bubble, for instance), only two bubble specifications have directly been examined in literature so far: a deterministic bubble, in the context of German hyperinflation analysed by Flood, Garber (1980), and an intrinsic bubble, in the context of US–stock prices investigated by Froot, Obstfeld (1991b). The former study was the first test for the existence of rational bubbles at all. Subsequently, more and more indirect test procedures were developed inspired by the volatility tests by Shiller (1979, 1981b) and LeRoy, Porter (1981). Almost a decade after the first, a second direct test (a test for intrinsic bubbles) was designed, influenced by the disappointing results of and the problems connected with indirect tests for rational bubbles.

4.2.1 Deterministic bubble in German hyperinflation

Flood, Garber (1980) investigate the 1920–1923 hyperinflation in Germany by using Cagan's hyperinflation model presented in section 3.3. As a bubble component they included a deterministic bubble within the general solution which can be motivated by the observation that in periods of hyperinflations the increase of prices seems to be highly explosive and self–fulfilling. A deterministic bubble tries to capture precisely this explosiveness[51].

Flood, Garber (1980) start with the following model specification describing the determination of the money process in face of hyperinflationary

[51]One must, however, be aware of the fact that deterministic rational bubbles can only exist for a bounded period of time since economic variables do not monotonically increase without end. In the case of hyperinflations the observation period is naturally bounded by a monetary reform which usually terminates a hyperinflationary processes. Nevertheless, one must consider the possible sensitivity of tests for deterministic rational bubbles concerning the observation period to be examined.

tendencies[52]

$$m_t - p_t = \gamma_0 + \gamma_1 E[\pi_{t+1} \mid I_t] \quad \gamma_1 < 0 \tag{4.22}$$

where m_t and p_t denote the logarithms of nominal money[53] and price, $E[\pi_{t+1} \mid I_t] = E[p_{t+1} \mid I_t] - p_t$ the rationally expected rate of inflation, and γ_0, γ_1 reaction coefficients[54]. Assuming an equilibrium on the money market, m_t equals money supply which is supposed to be exogenously determined and represents at the same time the monetary equilibrium. From equation (4.22) the following general solution can be derived

$$\begin{aligned} p_t &= \sum_{i=0}^{\infty} \alpha^i E[x_{t+i} \mid I_t] + \frac{1}{\alpha^t} A_0 \\ \text{with} \quad \alpha &= -\frac{\gamma_1}{1-\gamma_1} \\ x_{t+i} &= \frac{m_{t+i} - \gamma_0}{1-\gamma_1} \end{aligned} \tag{4.23}$$

Since $\gamma_1 < 0$, α is smaller than 1; hence, the forward solution is the appropriate fundamental solution of equation (4.23). $\frac{1}{\alpha^t} A_0$ characterizes a determinstic bubble whereas A_0 denotes an arbitrary constant[55].

Equivalently, the general solution can be expressed in terms of first differences, i.e. inflation rates and money growth rates, as

$$\begin{aligned} \pi_t &= \frac{1}{1-\gamma_1} \sum_{i=0}^{\infty} \alpha^i E[\mu_{t+i} \mid I_t] + \frac{1}{\alpha^t} \tilde{A}_0 \\ \text{with} \quad \mu_{t+i} &= m_{t+i} - m_{t+i-1} \\ \tilde{A}_0 &= (1-\alpha) A_0 \end{aligned} \tag{4.24}$$

where π_t stands for the inflation rate and μ_t for the money growth rate. Notice that the decisive assumption in the following testing procedure is

[52]For the sake of simplicity, we subsequently ignore all error terms which are originally included in Flood, Garber (1980).

[53]To be more precise, we should distinguish between money demand and money supply. Therefore, equation (4.22) actually determines the money demand which equals the money supply exogenously determined.

[54]For a more detailed discussion of equation (4.22) see section 3.3.

[55]Flood, Garber (1980) use a slightly different specification of the bubble term. We choose the above specification in accordance with our previous analysis. The results, however, are not affected.

that the process of money supply $\{m_t, t \in \mathcal{T}\}$ is determined by the government exogenously and thus does not respond to changes in inflation rates.

Flood, Garber (1980) conduct three different test procedures:

- *Joint estimation of the money demand and the inflation process.* Merely estimating a perfect foresight counterpart of equation (4.22) by using instrumental variable techniques, as proposed by McCallum (1976)[56] is inefficient since no information concerning the inflation rate is used. Taking this shortcoming of the McCallum–technique into account, Flood, Garber (1980) describe the inflation rate as being sufficiently representable by an autoregressive process of order k of the money growth rate μ plus a deterministic bubble component which mainly stems from equation (4.24)

  $$(4.25) \qquad \pi_t = \beta_0 + \beta_1\mu_{t-1} + \ldots + \beta_k\mu_{t-k} + \frac{1}{\alpha^t}(1-\alpha)A_0$$

  Substitution of equation (4.25) in (4.22) yields

  $$(4.26) \qquad \begin{aligned} m_t - p_t &= \gamma_0 + \gamma_1(\beta_0 + \beta_1\mu_{t-1} + \ldots + \beta_k\mu_{t-k} \\ &\quad + \frac{1}{\alpha^t}(1-\alpha)A_0) \end{aligned}$$

  Joint estimation of the equation (4.25) and equation (4.26), by using a non–linear seemingly unrelated regression (SUR) estimation procedure which simultaneously imposes non–linear cross–equation restrictions on the parameter[57], gives the first set of estimated parameters $(\hat{\gamma}_0, \hat{\gamma}_1, \hat{\beta}_0 \ldots \hat{\beta}_k, \hat{A}_0)$. Notice that α is a function of γ_1 and, therefore, not explicitly tested for significance.

- *Joint estimation of the money demand and the money supply.* Alternatively, Flood, Garber (1980) substitute equation (4.24), i.e. the

[56]The expected inflation rate is substituted by the actual inflation rate and its parameter is estimated by using various instrumental variables such as lagged growth rates of money. See McCallum (1976) and Engsted (1993a) for a description.

[57]This procedure can be seen as a generalization of the McCallum–technique and was first proposed by Singleton (1979). Flood, Garber (1980) use four lags of money growth within equation (4.25) for determining the inflation rate.

general solution, directly into equation (4.22). The money demand is then defined as

$$(4.27) \qquad m_t - p_t = \gamma_0 + \gamma_1\Big(\frac{1}{1-\gamma_1}\sum_{i=0}^{\infty}\alpha^i E[\mu_{t+i} \mid I_t] + \frac{1}{\alpha^t}(1-\alpha)A_0\Big)$$

Assuming money growth rates to follow an autoregressive process of order k[58],

$$(4.28) \qquad \mu_t = \phi_0 + \phi_1\mu_{t-1} + \phi_2\mu_{t-2} + \ldots + \phi_k\mu_{t-k} + \epsilon_t$$

the general solution of equation (4.22) can easily be calculated explicitly by using proposition 2.3 and 2.4. Joint estimation of equation (4.27) with repect to the best fitted AR–process for the money growth rate and equation (4.28) by again using a non–linear seemingly unrelated regression (SUR) model which simultaneously imposes non–linear cross equation restrictions on the parameters[59] gives the second set of estimated parameters $(\hat{\gamma}_0, \hat{\gamma}_1, \hat{\phi}_0 \ldots \hat{\phi}_k, \hat{A}_0)$.

- *Two step procedure.* The last procedure substitutes the parameter of equation (4.28), estimated by using standard techniques of time series analysis, into equation (4.24) and (4.27). The resulting equations were again estimated to jointly yield the remaining parameters $(\hat{\gamma}_0, \hat{\gamma}_1, \hat{A}_0)$ [60].

In their analysis of the German hyperinflation Flood, Garber (1980) use monthly data from July 1920 to June 1923, in particular the wholesale price index and the mid–month realization of the money supply, which was calculated by using an interpolation technique[61]. Additionally, different bubble periods are considered by including a dummy variable multiplicatively connected with the bubble term in equation (4.23). All three

[58] Flood, Garber (1980) apply standard techniques of time series analysis in order to identify the process of money growth rates. Univariate identification routines lead to the conclusion that μ_t can best be described by an AR(1)–process in first differences.

[59] This procedure was developed by Hansen, Sargent (1980).

[60] Although this procedure gives inefficient parameter estimates since cross equation restrictions are not used, it benefits from more data included because of independent estimation of μ_t.

[61] See Flood, Garber (1980) for a more detailed description.

test procedures do not find any significant estimate $\hat{A}_0$ at any reasonable significance level. Consequently, Flood, Garber (1980) conclude that during the German hyperinflation during 1920–1923 a deterministic bubble cannot be found in the data.

The following controversial issues concerning the Flood, Garber (1980) analysis can be made:

- First, the assumption of exogenous money supply may adversely affect the outcome. A model which admits a feedback mechanism from inflation to money supply probably comes closer to reality. Results different from those described above may be obtained.
- Second, the strong correlation between the estimates of α as a function of γ_1 and A_0 is obvious. The low significance of the bubble term then may be due to this correlation. In these circumstances, a likelihood-ratio test probably performs better than standard t-tests[62].

The subsequent study of Flood, Garber, Scott (1984) tries to capture some of these problems. Maintaining the assumption of exogenous money supply, Flood, Garber, Scott (1984) extend the analysis of hyperinflationary phenomena after I World War by independently conducting the first testing strategy of Flood, Garber (1980) additionally for Hungary and Poland. Moreover, they use likelihood ratio tests in addition to the standard t–tests in order to test for the significance of the estimated parameter values of the bubble term mitigating the problem of strong correlation between γ_1 and A_0. Finally, it is tested whether a bubble was simultaneously present in the three countries considered[63]. They apply a maximum likelihood estimation procedure when using the first estimation strategy of Flood,

[62]Interestingly, in all three test procedures parameter estimates for γ_1 are highly significant.

[63]This procedure was mainly motivated by the notion that by merely increasing the number of observations with respect to time, the asymptotic distribution of the parameter degenerates, due to the explosive bubble term. By adding 'parallel' rational bubbles (i.e. bubble processes which exist in different countries simultaneously), the problem is avoided and more data can be used. See Flood, Garber (1980) footnote 18 and Flood, Garber (1990).

Garber (1980), and find that, at least for the German case, 'there is significant improvement in the fit of the equations if we relax the restriction that $A_0 = 0$' (Flood, Garber, Scott (1984), p 336). This result may be interpreted as an evidence that a deterministic bubble existed in the German hyperinflation.

Finally, it should be noted that Casella (1989), unlike both Flood, Garber (1980) and Flood, Garber, Scott (1984), consider additionally endogenous money supply responses in face of increasing inflation rates. However, by using specification test procedures, rather than direct test procedures, she found no evidence for rational bubbles if money supply is assumed to be determined endogenously. In the case of exogenous money supply some evidence in favour of rational bubbles still prevails[64].

In sum, the evidence for existing rational bubbles in German hyperinflation still remains an open issue. In particular, joint estimation of the bubble parameter in direct test procedures and exogeneity or endogeneity of money supply may be seen as major problems.

4.2.2 Intrinsic bubbles on stock markets

After a decade of indirect tests, Froot, Obstfeld (1991b) set up a framework for another direct test for the presence of rational bubbles by introducing the concept of intrinsic bubbles[65]. The application of a direct test for the existence of intrinsic bubbles can be justified by the following empirical and statistical reasons. First, a test for intrinsic bubbles has at least the potential to explain at most a part of the empirically observed excess volatility on capital markets. Second, due to its non-linear nature, a direct test for intrinsic bubbles can picture the effect of an overreaction of stock prices to news (changes in fundamentals, for instance) often observed on stock

[64]Furthermore, Burmeister, Wall (1982, 1987) use Kalman filtering techniques to investigate the explanatory power of rational bubbles within the period of German hyperinflation. Both papers confirm the possible existence of rational bubbles, the former relying on exogenous, the latter on endogenous money supply.

[65]Notice that Froot, Obstfeld (1991b) use the special intrinsic bubble G4 of table 2.8.

markets[66]. Thirdly, unlike other bubble specifications[67], intrinsic bubbles are econometrically testable, as shown by Froot, Obstfeld (1991b). Other bubble specifications contain extremely explosive components which are often not observed in economic data or remain untestable, due to included probabilities, like the bursting bubble. Due to its dependence on the fundamental process, intrinsic bubbles, however, avoid the extreme explosiveness common to other bubble specifications[68]. And finally, when accepting the principle of parsimonious model building[69], a general solution, containing an intrinsic bubble specification, depends only on the fundamental process, avoiding the inclusion of other extraneous variables.

The starting point of the analysis of intrinsic bubbles is the well known equation (2.4) for the stock market, supplemented by a stochastic error term u_t denoting a predictable one–period excess return

$$p_t = \exp(-r)[E[p_{t+1} \mid I_t] + d_t] + \exp(-r)u_t$$

whereas p_t and d_t denote the stock price and the fundamental process connected with the stock in question. r stands for the real rate of interest assumed to be constant over time. Notice that $\alpha = \exp(-r)$ in this case represents a continuous time discount factor. Additionally, the fundamental process is assumed to follow a geometric random walk, i.e. the log of the fundamentals can be described as

$$\ln d_t = \ln d_{t-1} + \mu + \zeta_t, \quad \zeta_t \sim IIN(0, \sigma^2)$$

By applying the different solution procedures to the case of intrinsic bubbles with fundamentals being characterized by a geometric random walk, as presented in chapter 2, we get the following general solution

$$p_t = c_0 d_t + c_1 d_t^{\lambda} + v_t \tag{4.29}$$

---

[66]See section 3.1.2.

[67]See table 2.8.

[68]Naturally, intrinsic bubbles also show an explosive behaviour by construction, like all bubble specification. Their dependence on the fundamental process, however, dampens their explosiveness.

[69]See McCallum (1983).

The first term on the left hand side refers to the fundamental solution[70], the second represents the bubble solution. v_t is the present value of all expected future error terms u_t, i.e. $v_t = \sum_{i=0}^{\infty} \exp(-r(i+1))E[u_{t+i} \mid I_t]$. Recall that the coefficient c_0 denotes the parameter derived theoretically for the fundamental solutionif the fundamental process can be described by a geometric random walk, i.e. $c_0 = \left(\exp(r) - \exp(\mu + \frac{\sigma^2}{2})\right)^{-1}$. λ can be calculated by solving the characteristic equation of all corresponding processes depending on the fundamentals d_t. Froot, Obstfeld (1991b) consider only the corresponding positive root, i.e. $\lambda > 0$. Notice that for convergence of the fundamental solution we must impose the restriction $r > \mu + \frac{\sigma^2}{2}$[71]. Taking into account this restriction, it is easily shown that $\lambda > 1$, establishing the effect of overreaction to changes in fundamentals which intrinsic bubbles intent to describe .

In order to avoid the problem of multicollinearity among the regressors, Froot, Obstfeld (1991b) divide equation (4.29) by d_t and get the following final regression equation for the price–fundamental ratio

$$\frac{p_t}{d_t} = c_0 + c_1 d_t^{\lambda-1} + \eta_t \quad \text{with} \quad \eta_t = \frac{v_t}{d_t} \tag{4.30}$$

Furthermore, Froot, Obstfeld (1991b) show that under certain assumptions concerning the residuals η_t, the t–statistic of $\hat{c}_1$ is asymptotically distributed standard–normally, despite the presence of an explosive regressor. These decisive, partly controversial, assumptions are

- The residuals η_t are distributed normally and identically with zero unconditional mean. They can, however, imply heteroskedasticity and autocorrelation.

- The residuals η_t are independent from fundamentals' innovation ζ_t in all leads and lags. This means that the residuals do not contain any information which can be drawn from fundamentals[72].

[70]Since $\exp(-r) < 1$ for $r > 0$, the parameter α lies within the required interval for applying the forward solution.

[71]See proposition 2.10.

[72]Naturally, they can contain information concerning other variables like time–varying discount rates or risk premia, among others.

Froot, Obstfeld (1991b) apply two estimation procedures. First, they estimate equation (4.30) by non–linear OLS correcting residuals for heteroskedasticity using Newey, West (1987) covariance matrix correction procedure for serial correlation of unknown form. Second, maximum likelihood estimation techniques are applied assuming the residuals to be representable by an AR(1)–process. Additionally, they conduct a restricted regression by forcing the parameter λ to be equal to the theoretically warranted value[73], again using non–linear least squares and maximum likelihood estimation techniques.

For statistical inference Froot, Obstfeld (1991b) argue that besides common t–tests in order to test for the significance of the bubble term, an F–test, which coincides with a likelihood–ratio test, should be applied since the derivatives of the parameter λ and c_1 are highly, positively correlated.

Froot, Obstfeld (1991b) examine the period 1900–1988 using Standard & Poor stock price and dividend indexes. As endogenous process they consider the January–average of nominal stock prices of each year deflated by the January producer price index (PPI). Dividends are nominal averages for each year deflated by the year–average PPI[74].

The results of their analysis can be summarized as follows: The estimated parameters of equation (4.30) are all positive (especially for c_1 and λ), show economically reasonable values and appear to be very similar to their corresponding counterparts, theoretically derived. By applying common t–tests, the decisive coefficient of the bubble term $\hat{c}_1$ is significant in the case of restricted regression but insignificant in the case of unrestricted regression. However, $\hat{c}_1$ gains significance in the unrestricted case when an F–test is applied. Mainly based on the results of the LR–test and of the restricted regression, Froot, Obstfeld (1991b) conclude that intrinsic bubbles are likely to be present on the US–stock market during this century and are able to explain at least a part of the excess returns connected with stocks. Discussing the Froot, Obstfeld (1991b) procedure for testing for the existence of intrinsic bubbles the following points must be emphasized:

[73] λ was calculated by using sample mean and sample variance of the dividend process and average annual return on stocks.

[74] Some results of their analysis may be sensible to a different timing.

- The direct test procedure presented below can only reasonably be conducted for fundamentals which can be described by a geometric random walk[75]. A more complex ARIMA–process has at least the potential to explain the fundamental process better than simple non–stationary geometric random walk processes. The test may, therefore, underly a severe oversimplification in the case of the fundamentals.

- Froot, Obstfeld (1991b) use annual data of US–stock market from 1900 to 1988. A more convincing analysis would apply to a lower data frequency if the relevance of rational bubbles is to be investigated.

- The main problem in directly testing for the relevance of bubbles is the inclusion of non–stationary regressors (here, the fundamentals follow a geometric random walk) and of the explosiveness of the bubble term. Therefore, the distributional properties of the corresponding t–statistics are unknown. Here, Froot, Obstfeld (1991b) show by means of a Monte–Carlo study the near–normality of $t_{\hat{c}_1}$.

Nevertheless, the Froot, Obstfeld (1991b) direct test for intrinsic bubbles appears to provide an interesting procedure worthy of further application on other stock markets, however, slightly modified concerning the data frequency. Bruns (1994) was the first to apply the Froot, Obstfeld (1991b) analysis to the German Stock Market. He uses annual data (1951–1993) and monthly data (01:1978–12:1993) on stock prices and dividend indexes and does not find any significant impact of intrinsic bubbles. Moreover, most of the estimates fail to coincide with their theoretically expected values[76].

---

[75]If the fundamental process follows a geometric random walk, the hypothesis that the fundamental process follows a random walk with drift, cannot be rejected for most of the fundamental processes, either. Therefore, a non–stationary random walk with drift can also be used in the analysis resulting in slightly different non–linear relationships to be tested for. However, the analysis is sensible with respect to other, possibly stationary, fundamental processes because the corresponding general solution differs in this case significantly from the one derived above.

[76]In particular, λ turns out to be negative.

In order to justify the reapplication of the Froot, Obstfeld (1991b) test for the German Stock Market, some shortcomings, connected with Bruns' analysis, should already be mentioned here. First, no LR–test was carried out in order to test for the significance of the bubble term. Recall that in the unrestricted case only the LR–test indicates a significant influence of the bubble component on stock prices. Second, no preliminary investigation was conducted for gaining evidence of the geometric random walk hypothesis for the underlying fundamentals. And finally, in accordance with Froot, Obstfeld (1991b), Bruns uses only an index of the entire stock market. A more comprehensive analysis may additionally consider sector indexes and single stocks.

These remarks may support the reapplication of the intrinsic bubble test procedure for the German Stock Market which will be carried out in the next chapter. The extensions, based on the preceding remarks, will be discussed there in greater detail.

4.2.3 An econometric caveat

Before finally assessing the relevance of direct test procedures, a central problem, mainly associated with direct test procedures, must be discussed here, i.e. the *observational equivalence* [77] between the bubble term occurring in the general solution of equation (2.4) and arbitrary other factors which are not observed and are, therefore, not integrated into the testing procedure by the econometrician.

For illustrating this problem we first reformulate the general description of a bursting bubble in terms of initial value of the bubble, i.e. b_0. Omitting all error terms in (2.46), we get

$$b_t = \begin{cases} \frac{1}{(\pi\alpha)^t} b_0 & \text{with probability } \pi^t \\ 0 & \text{with probability } 1 - \pi^t \end{cases}$$

where $b_0 > 0$ denotes the starting value of the bursting bubble at time $t = 0$. Notice that the probability of the bursting bubble existing in

[77] These points were first made by Flood, Garber (1980) within a deterministic bubble framework and by Hamilton, Whiteman (1985) and Hamilton (1986) for the case of bursting bubbles.

period t, i.e. π^t, decreases as time goes by. Suppose[78] now fundamentals are constant at level $\bar{x}$. Then, the corresponding general solution including the bursting bubble description is easily obtained.

$$(4.31) \qquad y_t = \begin{cases} \frac{1}{1-\alpha}\bar{x} + \frac{1}{(\pi\alpha)^t} b_0 & \text{with probability } \pi^t \\ \frac{1}{1-\alpha}\bar{x} & \text{with probability } 1 - \pi^t \end{cases}$$

In order to show the occurrence of observational equivalence between this general solution including a bursting bubble and solely a fundamental solution containing an additional unobserved component, an arbitrary process n_t is introduced which is constant at level $\bar{n}$, but possibly crashes to zero every period from $t = 0$ to T with probability π.

$$n_t = \begin{cases} \bar{n} & \text{with probability } \pi^t \\ 0 & \text{with probability } 1 - \pi^t \end{cases} \qquad \text{for } t < T$$

After period T the process simply remains constant at the level realized in period T. For $t < T$ the fundamental solution is easily established as

$$\begin{aligned} y_t &= \frac{1}{1-\alpha}\bar{x} + \sum_{i=0}^{T-t-1} \alpha^i \pi^{i+1} \bar{n} + \sum_{i=T-t}^{\infty} \alpha^i \pi^{T-t} \bar{n} \\ &= \frac{1}{1-\alpha}\bar{x} + \frac{1}{1-\alpha\pi}\bar{n} - (\alpha\pi)^{T-t}\frac{1}{1-\alpha\pi}\bar{n} + (\alpha\pi)^{T-t}\frac{1}{1-\alpha}\bar{n} \\ (4.32) \qquad &= \frac{1}{1-\alpha}\bar{x} + \frac{1}{1-\alpha\pi}\bar{n} + \frac{1}{(\alpha\pi)^t} \underbrace{\left[\frac{1}{1-\alpha} - \frac{1}{1-\alpha\pi}\right](\alpha\pi)^T \bar{n}}_{=b_0} \end{aligned}$$

Notice that the last term of equation (4.32) can be identified with the bubble solution in equation (4.31)[79]. Hence, both the general solution

---

[78]This example mainly follows Hamilton (1986).

[79]Two special cases often encountered in literature can be derived from equation (4.32). First, suppose only at time T a crash will occur inspired, for instance, by the uncertain outcome of an election. Then, the fundamental solution for $t < T$ reduces to $y_t = \frac{1}{1-\alpha}\bar{x} + \frac{1}{(\alpha\pi)^t} \underbrace{\frac{(\alpha)^T}{1-\alpha}\pi\bar{n}}_{=b_0}$. Notice that the last term of the fundamental solution can be interpreted as a bursting bubble component. Second, suppose the increase of fundamentals by the amount of $\bar{n}$ occurs with certainty at T, i.e. $\pi = 1$. Then, the equivalence of both a general solution containing a (deterministic) bubble component and the fundamental solution including the component $\bar{n}$ is obvious. Notice that in the deterministic case the equivalence is due to the anticipation effect of rational expectations. A certain event, expected to occur at time T, has an effect on the time path of the endogenous variable y_t previous to time T.

containing a bubble term and a fundamental solution including an unobserved variable are observationally equivalent for $t < T$. The fundamental solution for $t > T$ equals

$$y_t = \begin{cases} \frac{1}{1-\alpha}(\bar{x} + \bar{n}) & \text{with probability } \pi^T \\ \frac{1}{1-\alpha}\bar{x} & \text{with probability } 1 - \pi^T \end{cases} \quad \text{for } t > T$$

The interpretation of the variable n_t is twofold:

- First, n_t may stand for a change in the fundamental process itself, only observed by the market agents, but not considered by the econometrician. Assuming $\bar{n} > 0$, the corresponding increase of fundamentals is expected to vanish in every period from $t = 0$ to T with a probability equal to $1 - \pi$. After period T the fundamentals retain their value achieved in T. This *change in the fundamental process itself* [80] may be due, for instance, to better economic prospects of the corresponding company which are, however, likely to be revised every period with probability $1 - \pi$. Hence, a somewhat fragile economic situation, not taken into account by the econometrician, may induce a bubble–like behaviour on stock markets.
- Moreover, n_t may represent any other *variable not observed by the econometrician*[81] while market agents think this variable as to influence the motion of the endogenous variable. The expectation of a reform in taxing dividend payments which have a negative impact on the rate of return, or other arbitrary factors, like the often cited result of a US–super bowl football final, may serve as examples of these variables. Notice that here again a sort of extrinsity enters the model. The similarity to the concept of sunspots or extrinsic bubbles is obvious.

A graphical representation may help understand the effect of n_t on the endogenous variable. Figure 4.2 shows the time path of the fundamental

[80] This effect was emphasized by Hamilton (1986). There are other approaches mainly corresponding to exchange rates where a stochastic regime or process switching can produce bubble–like behaviour. See, for instance, Flood, Garber (1983, 1991) and Froot, Obstfeld (1991a, 1991c).

[81] This interpretation is due to Flood, Garber (1980).

solution (4.32) between $t = 0$ and T for $\bar{n} > 0$. Notice that in every period the bubble may burst, forcing the endogenous variable to return to $\frac{1}{1-\alpha}\bar{x}$. Notice further that an immediate increase in the endogenous

Figure 4.2: Stylized time path of equation (4.32)

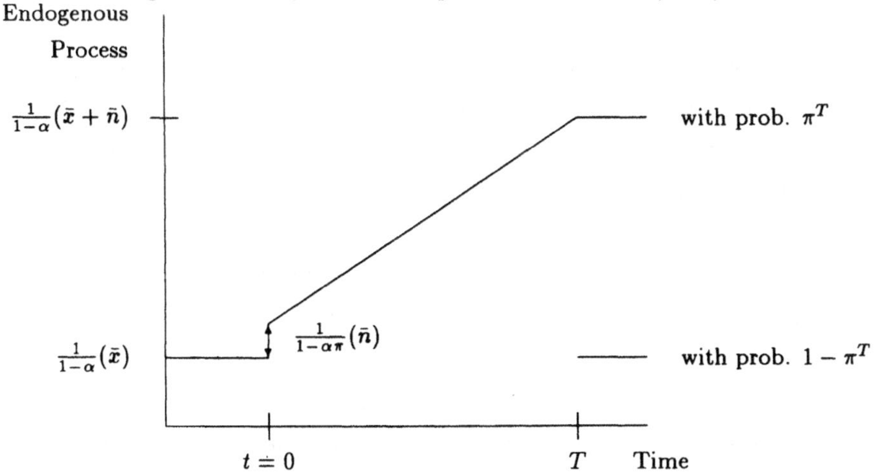

variable in period $t = 0$ occurs if the variable n_t is expected to influence y_t. Subsequently, when n_t does not crash to zero, the time path of y_t grows exponentially until $\frac{1}{1-\alpha}/(\bar{x}+\bar{n})$ is reached.

The argument against directly testing for rational bubbles is clear. If the bubble component is indistinguishable from observationally equivalent variables, a direct testing strategy may falsly confirm the presence of rational bubbles while other factors have induced the endogenous variable to deviate from the corresponding fundamental path. However, in order to justify the application of direct test procedures, two points should be mentioned:

- First, this objection to direct testing generally meets every econometric model building. The problem of omitted variables has often been addressed in econometric textbooks, and the related problems are well known. The consequence in face of this problem is to check carefully the underlying economic model, i.e. we must ask whether the fundamentals, including their possible pattern, are correctly identified and whether all variables, important for explaining the behaviour of stock prices, are considered in the model.

- Second, the time path of the endogenous process, including an observationally equivalent component pictured above, shows the behaviour commonly ascribed to rational bubbles. This time path relates to a situation where a positive effect, $\bar{n} > 0$, is likely to vanish within a certain time span from $t = 0$ to T. Contrarily, assuming a negative effect of the observationally equivalent component, $\bar{n} < 0$, a behaviour of the endogenous process converse to the time path seen above would be produced. Under the condition that both situations, i.e. a positive and a negative effect, are equally likely, we should observe both time path patterns in actual data series. However, it turns out that the endogenous process does not increase so sharply as it abruptly falls down at times of crashes. Why, then, should these components, which are observationally equivalent to rational bubbles, occur merely as positive effects? Consequently, it must be doubted that these phenomena produce only a strong decrease in the time series involved when strong increases are supposed to be equally likely.

To conclude, econometric modelling must always be aware of omitted variables which are important for the problem considered. When testing for the presence of rational bubbles, special attention must be paid in determining the underlying fundamentals and other variables which may influence the endogenous process.

4.2.4 Final assessment of direct tests

Besides the possibility of a misspecification of the underlying model already discussed at the end of the chapter 3, emphasizing possible alternatives based mainly on time–varying discount rates (parameter α), risk–averse individuals and irrational behaviour in connection with heterogenous spread of information, some econometric problems connected with direct test procedures became obvious in the preceding subsection. What else can be said about direct test procedures in order to reach a final assessment?

Unlike indirect tests, a direct test for rational bubbles avoids the problem of interpreting a rejection of the hypothesis of no rational bubbles. Most indirect test procedures provide no evidence of the existence of rational bubbles if the hypothesis of no bubbles is rejected. The deviation of asset prices from its present value can be explained by any other economically reasonable source, e.g. time–varying discount rates, risk premia or fads. The approach presented here focuses directly on the presence of rational bubbles. If the corresponding coefficients are significant, rational bubbles have an explanatory impact on the endogenous variable considered. On the other side, insignificant coefficients of the bubble component do not mean that rational bubbles are not at all present. Due to the multiplicity of the bubble solution, other kinds of rational bubbles, which were not directly tested for, still remain possible. All in all, although both direct and indirect procedures imply serious problems, direct test approaches appear to be more convincing than indirect ones, basically because of greater interpretability of the results. Moreover, the results of chapter 2, where the set of bubbles has been enlarged significantly, encourage to confirm the empirical relevance of these processes rather directly than indirectly.

However, direct test procedures are only recommended if the model considered is unlikely to be misspecified, the underlying fundamentals are well defined such that process switching is easily observed by the researcher, and the probability of unobserved variables, influencing the endogenous variable, is low. Comparing the three main applications in the bubble theory, testing for rational bubbles can reasonably undertaken only for the case of capital markets. For instance, in comparison to the foreign exchange rate framework, the underlying fundamentals related to stock prices are easier to determine than those related to exchange rates. This is mainly due to the fact that no commonly accepted model of exchange rate determination, which is necessary to identify the underlying fundamentals, exists in the case of foreign exchange markets. That provides the main reason to basically concentrate on capital markets for the following test strategy applied to the German Stock Market.

Chapter 5

On the Explanatory Power of Rational Bubbles on the German Stock Market

In order to test for rational bubbles on the German Stock Market directly, two different bubble specifications will be considered. First, the presence of a Markovian bubble in the sense of a deterministic or bursting bubble will be examined and, second, the existence of intrinsic bubbles is tested for. Concerning the latter bubble specification, two distinct characterizations, a temporary and a permanent, will be investigated. Moreover, it will turn out that some preliminary investigations based on unit roots tests should reasonably be undertaken. Therefore, a special digression has been added at the end of this chapter describing the most common techniques for testing for the degree of integration, applying these techniques to the processes considered and discussing the results. However, before turning to the precise description of direct bubble and integration test procedures, the data used in the analysis below need to be specified.

5.1 Data

Unlike the convenience of many empirical investigations for rational bubbles on capital markets, we are convinced that despite their possibly lower power in statistical inference[1], the use of short period data (quarter, monthly, daily) is more appropriate than using annual data since the occurrence of rational bubbles capturing the idea of self-fulfilling expectations is an inherently short term phenomenon. The reasonability of concentrating on the short run may further be justified by the fact that when using annual data, the period covered by the data in order to get a sufficient sample size, has to be large[2]. The probability, however, that structural breaks or changes occur increases seriously when the period considered becomes larger and larger. A severe trade-off between possibly including structural breaks in order to gain a sufficient sample size in the case of large observation periods and obtaining possibly problematic parameter estimates in the case of smaller observation periods arises.

We use both annual data and quarter or monthly data, if possible. The use of annual data is reasonable since data on dividends, representing apparently the most logical description of fundamental processes underlying the stock market, are available only once a year – at least if German data are used. If we want to work with short period data, proxy variables which are obtainable for shorter time spans and which can simultaneously be thought of as influencing stock prices[3], have to be introduced. These proxy variables are presented below in greater detail.

As mentioned above, structural breaks can be present in data covering long periods of time. Especially for the German Stock Market, two structural breaks in the last decades should be noted. First, every time series on German economic data is bounded by World War II. Accordingly, data on capital markets usually start at the end of the 1950s leading unfortunately to a small sample size if annual data are used. Second, in 1976 a reform of taxing dividends was established in Germany. Therefore, time series of both dividends and stock prices containing periods around 1976 are likely to be biased. Consequently, when using annual data, we are forced to

---

[1]See Shiller, Perron (1985).

[2]Froot, Obstfeld (1991b), for instance, use US-data from 1900–1988.

[3]See Campbell, Shiller (1988) for a similar approach. They argue that earnings, for instance, can reasonably be used as fundamentals, and show that long moving averages of earnings indeed have a significant impact on stock prices.

apply data from 1960[4] being aware of the fact that a structural break can be present in the data. When using quarter or monthly data, we start in 1978 avoiding a possible bias, due to the tax reform.

As *endogenous processes* three types of data are considered:

- First, *stock price indexes* (In) play a prominent role in every empirical work on capital markets. We use three stock price indexes: the DAFOX provided by the 'Karlsruher Kapitalmarktdatenbank'[5], an index computed by the 'Frankfurter Allgemeine Zeitung' [FAZ–Index] representing one of the oldest indexes continuously published in Germany since 1958 and an index composed by the 'Statistisches Bundesamt' [StBA–Index][6]. A short characterization of the indexes considered is summarized in table 5.1. Notice the different correction procedures applied when calculating the indexes.

Table 5.1: Stock price indexes

	DAFOX	FAZ–Index	StBA–Index
No. of Stocks	317*	100**	288***
Weights	Nominal capital	Nominal capital	Nominal capital
Correction	Capital increase Dividend payments	Capital increase	Capital increase
Periodicity	Daily	Daily	Monthly averages
Source	Karlsruher Datenbank	Karlsruher Datenbank	Statistisches Bundesamt

Remarks:
*: variable number of stocks, cited value corresponds to 12:91
**: fixed number of stocks
***: variable number of stocks, cited value corresponds to 12:89

- Second, average monthly data on *sector indexes* (Se) are considered in the analysis below. These sector indexes represent subclasses of stocks which refer to the same industry and are provided by the 'Statistisches Bundesamt'. They are calculated in the same way as

[4]Data from 1958 are available only for the FAZ–Index.

[5]For a detailed description and calculation of the index see Göppl, Schütz (1994).

[6]Unfortunately, the DAX the most popular stock price index in Germany, starts first in 1988.

the StBA–Index. The sector indexes used in our analysis are industry, investment, consumption, energy, chemical industry, mechanical engineering, car industry, warehouses, banking, and insurance, covering the most important sectors in Germany.

- Finally, the time path of seven single *stocks* (St) quoted at the Frankfurt stock exchange are tested for the presence of rational bubbles. Since reasonable proxy variables, which serve as fundamentals besides their own dividends, should be obtainable for these stocks, the choice of suitable candidates is limited[7]. Table 5.2 shows the chosen stocks and relates them to the corresponding sectors. The time

Table 5.2: Single stocks

	Corresponds to
Deutsche Babcock AG	Mechanical engineering
BASF AG	Chemical industry
Bayer AG	Chemical industry
BMW AG	Car industry
Phillip Holzmann AG	Construction
Karstadt AG	Consumption
Mannesmann AG	Mechanical engineering

series of all single stocks are again provided by the 'Karlsruher Kapitalmarktdatenbank' on a daily basis and delivered already corrected for dividend payments and capital increases.

The *fundamental processes* used in the analysis below can be divided into three different groups:

- First, *dividend payments* (Div) per year are considered as fundamental processes. We use both an index of dividends provided by the 'Statistisches Bundesamt' and the time series of dividends connected

[7]For instance, all stocks of service companies (banking and insurance companies above all) are not practicable since proxy fundamentals for those stocks are difficult to find. Moreover, some stocks cannot uniquely be assigned to a special sector like Siemens AG since the company's activity is related to different sectors.

Table 5.3: Combinations of endogenous and fundamental processes

	In1	In2	In3	Se1	Se2	Se3	Se4	Se5	Se6	Se7	Se8	Se9	Se10	St1	St2	St3	St4	St5	St6	St7
Div1	1	2	3																	
Div2														4						
Div3															5					
Div4																6				
Div5																	7			
Div6																		8		
Div7																			9	
Div8																				10
Na1	11	14	17																	
Na2	12	15	18																	
Na3	13	16	19																	
Na4				20	21			23	24	25										
Na5						22					26									
Na6												27	28							
Pi1				29																
Pi2							39													
Pi3				30																
Pi4								40							55	58				
Pi5					33															
Pi6									43					52						70
Pi7										46							61			
Pi8						36													67	
Pi9											49							64		
Io1				31																
Io2								41							56	59				
Io3					34															
Io4									44					53						71
Io5										47							62			
Io6						37					50								68	
Io7																		65		
It1				32																
It2								42							57	60				
It3					35															
It4									45					54						72
It5										48							63			
It6						38					51								69	
It7																		66		

Abbreviations

In	Stock indexes	St	Stocks	Na	National accounting data
In1	DAFOX	St1	Deutsche Babcock AG	Na1	National domestic product
In2	FAZ	St2	BASF AG	Na2	Gross capital income
In3	StBA	St3	Bayer AG	Na3	Gross value added (GVA)
Se	Sector indexes	St4	BMW AG	Na4	GVA manufactoring industry
Se1	Industry	St5	Phillip Holzmann AG	Na5	GVA trade and traffic
Se2	Investment	St6	Karstadt AG	Na6	GVA services
Se3	Consumption	St7	Mannesmann AG		
Se4	Energy	Div	Dividends		
Se5	Chemical industry	Div1	Dividend index		
Se6	Mechincal engineering	Div2	Div-Babcock		
Se7	Car industry	Div3	Div-BASF		
Se8	Warehouses	Div4	Div-Bayer		
Se9	Banking	Div5	Div-BMW		
Se10	Insurance	Div6	Div-Holzmann		
Pi	Production index	Div7	Div-Karstadt		
Pi1	Industry	Div8	Div-Mannesmann		
Pi2	Electricity and gas	Io	Index of order entries	It	Index of turnover
Pi3	Manufactoring industry	Io1	Manufactoring industry	It1	Manifacturing industry
Pi4	Chemical industry	Io2	Chemical industry	It2	Chemical industry
Pi5	Investment goods	Io3	Investment goods	It3	Investment goods
Pi6	Mechanical engineering	Io4	Mechanical engineering	It4	Mechanical engineering
Pi7	Car industry	Io5	Car industry	It5	Car industry
Pi8	Consumption goods	Io6	Consumption goods	It6	Consumption goods
Pi9	Construction	Io7	Construction	It7	Construction

with every stock considered above. The latter dividend time series are obtained from the 'Karlsruher Kapitalmarktdatenbank'.

- Second, *national accounting data* (Na) as proxies for the underlying fundamentals are supposed to have a statistically significant influence on appropriately chosen endogenous processes. We consider quarterly data on gross domestic income, gross capital income[8] and the gross value added for different sectors (manufacturing industry, trade and traffic, services). This set of fundamentals is delivered by the 'Statistisches Bundesamt' already corrected for seasonal influences.

- Finally, we use other *proxy variables* like adjusted production indexes (Pi) of sectors under consideration, the index of order entries (Io) or the index of turnovers (It) as an approximation of the underlying fundamentals. These time series are provided by the 'Statistisches Bundesamt' already deflated by the producer price index (PPI) and corrected for seasonal influences. We use both quarterly and monthly data of the corresponding time series.

Notice that all time series used in the following analysis are real data deflated by the producer price index (PPI)[9] provided by the 'Statistisches Bundesamt' except for proxy data like production indexes, indexes of order entries and indexes of turnovers already deflated by the 'Statistisches Bundesamt'. Table 5.3 summarizes all data used in our analysis, endogenous as well as fundamental variables, as presented above. Reasonable combinations of endogenous and fundamental variables are indicated by numbers. These combinations are called sets representing the basis of the following analysis in the sense that every set is tested for the presence of rational bubbles. In sum, 72 sets will be examined. In order to structure the sets defined in table 5.3, we construct classes combining sets of similar nature and timing. Table 5.4 gives an overview of the six different classes under investigation and describes their common nature.

[8]More precisely, this national accounting data category comprises the gross income on entrepreneurial activity and investment.

[9]We also tried time series deflated by the consumer price index (CPI), however, no significant difference was obtained in most regression results.

Table 5.4: Classes of sets

	Stock price indexes Source: KKMDB and StBA	Sector indexes Source: StBA	Single stocks Source: KKMDB
Dividends Source: KKMDB, StBA	**Class 1** Set 1-3 1960-1994 annual NOBS: 35		**Class 2** Set 4-10 1960-1994 annual NOBS: 35
National accounting data Source: StBA	**Class 3** Set 11-19 1:78-4:94 quarterly NOBS: 68	**Class 4** Set 20-28 1:78-4:94 quarterly NOBS: 68	
Proxies Source: StBA		**Class 5** Set 29-51 1:78-4:94 quarterly NOBS: 68 | 01:78-12:94 monthly NOBS: 204	**Class 6** Set 52-72 1:78-4:94 quarterly NOBS: 68 | 01:78-12:94 monthly NOBS: 204

Remarks: KKMDB = Karlsruher Kapitalmarktdatenbank; StBA = Statistisches Bundesamt; NOBS = number of observations

5.2 Direct test for rational bubbles

Based on the fundamental solutions derived in section 2.3 and on the theoretically possible bubble solutions (Markovian, intrinsic and extrinsic) displayed in table 2.8 of section 2.4, different combinations of fundamental and bubble solutions may be considered as a basis for testing for the existence of rational bubbles. However, it turns out that many bubble specifications lack a convincing economic interpretation or imply essential estimation difficulties making the corresponding combinations less favourable to be tested for[10]. Here, we choose three bubble specifications which are economically reasonable and econometrically tractable, i.e. a Markovian bubble in the sense of a deterministic or a bursting bubble (B2 in table 2.8 of section 2.4), an intrinsic bubble for fundamentals following a random walk (RW in table 2.8 of section 2.4) and the geometric intrinsic bubble II (G4 in table 2.8 of section 2.4)[11].

[10]For instance, bubble processes containing an exponential term prove to be difficult to estimate. Moreover, determining extraneous processes in the case of extrinsic bubbles seems to be rather complex.

[11]Notice that this is the Froot, Obstfeld (1991b) bubble.

Before describing the testing procedure more precisely, a few remarks concerning the timing of the bubble itself proves to be interesting. As it is easily seen, due to the very nature of the bubble processes mentioned above, the *first two bubble specifications* cannot be tested for their permanent presence. The inclusion of a time driven exponential term makes the endogenous process permanently expanding, a phenomenon not reflected by the data. Hence, only a temporary existence of these bubble specifications appears to be reasonable. This can be confirmed by looking at the graphical representation of most time paths involved. Figure 5.1 displays the time paths of the sector index of the entire industry as the endogenous process and the index of order entries of the entire industry as the fundamental process (set 31) in a twoscale graph. It becomes obvious that

Figure 5.1: Time paths of set 31

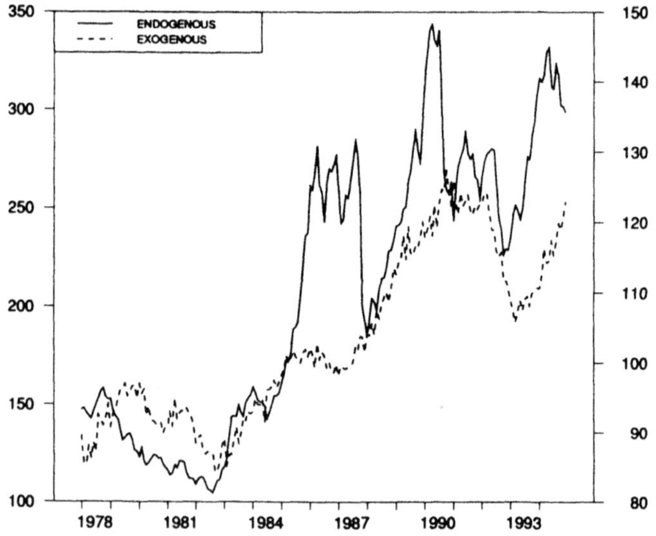

both processes develop very similarly (almost parallely) confirming the validity of simple present value models. However, between 1984 and 1987 the time paths appear to diverge in the sense that the endogenous process exceeds the fundamental process extensively. This apparently exponential deviation of the endogenous process from its fundamentals is terminated at the end of 1987, a date which is known for the occurrence of a dramatic crash in stock prices. A similar divergence of endogenous and exogenous process can be observed in 1989 and 1990 but the following crash appears to be smaller. These two phenomena can be interpreted as the presence of rational bubbles and supports the assumption of their temporary existence.

However, no evidence is provided as to which special kind of rational bubble may be present. The existence of Markovian as well as intrinsic bubbles is equally likely since both are able to explain explosive behaviour. Consequently, it seems to be reasonable to test for the temporary existence of both bubble specifications described above. Two reasonable periods for testing for the temporary existence of rational bubbles will be considered here:

- First, the period before October 1987 is examined, representing the time before the well known crash on the stock markets. We start the analysis of exploding bubbles in January 1980 in order to cover a long period and to obtain the 'true' starting point.

- Second, the same behavior of the stock prices is observed before September 1990 which can be regarded as a second crash on the stock markets within the last decades. We let the bubble start right after the occurrence of the first crash in November 1987.

Additionally, by inspection of figure 5.1, it turns out that, at least within the first bubble period considered, the bubble remains present on the stock market for a while without decreasing or increasing any further. Therefore, after the growth period an additional dummy variable is introduced to capture this 'duration period'.

Concerning the starting point and stopping time characterizing the timing of the bubble period, we apply an optimization procedure to determine the precise period in which these bubbles are likely to exist: An appropriate regression is run for different starting points beginning in January 1980 (November 1987). For each starting point different stopping times are considered. Then, we consider that combination of starting point and stopping time as 'optimal' which yields the best fit measured by the coefficient of determination. Figure 5.2 clarifies the optimization procedure

Figure 5.2: Stylized bubble period

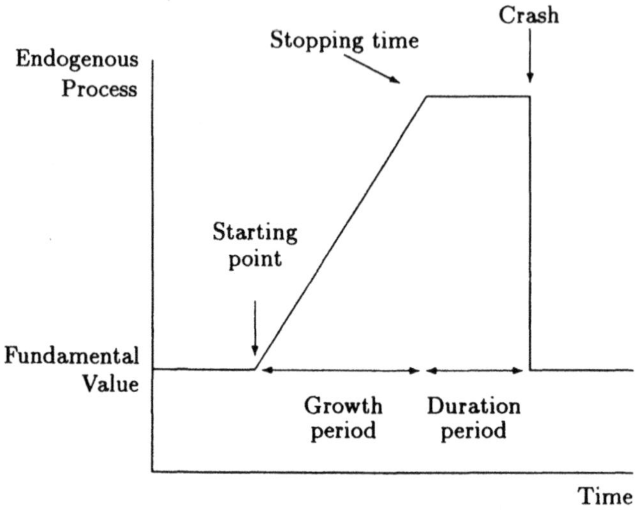

graphically. Notice that starting point and stopping time vary within the interval January 1980 (November 1987) to September 1987 (August 1990).

Finally, unlike the first two bubble specifications, the *third bubble type* will form the basis of testing for the permanent presence of intrinsic bubbles in the data. The lack of a time driven coefficient enables this application. Figure 5.3 summarizes the combinations of fundamental processes and bubble solutions to be examined. In the following we provide a deeper description of the testing strategy of the three bubble specifications and discuss their results.

5.2.1 Temporary Markovian bubbles

Based on the results of section 2.4, a Markovian bubble in the sense of a deterministic bubble (B2 in table 2.8 of section 2.4)[12] can be described as

$$b_t = \frac{1}{\alpha^t} b, \; b > 0 \tag{5.1}$$

[12] If we speak of a Markovian bubble here, we use the description of a deterministic bubble, bearing in mind that the time path behaviour described above actually concerns a bursting bubble which is equivalent to a deterministic bubble, if growing, except for a probabilistic term.

Figure 5.3: Bubble specifications to be tested

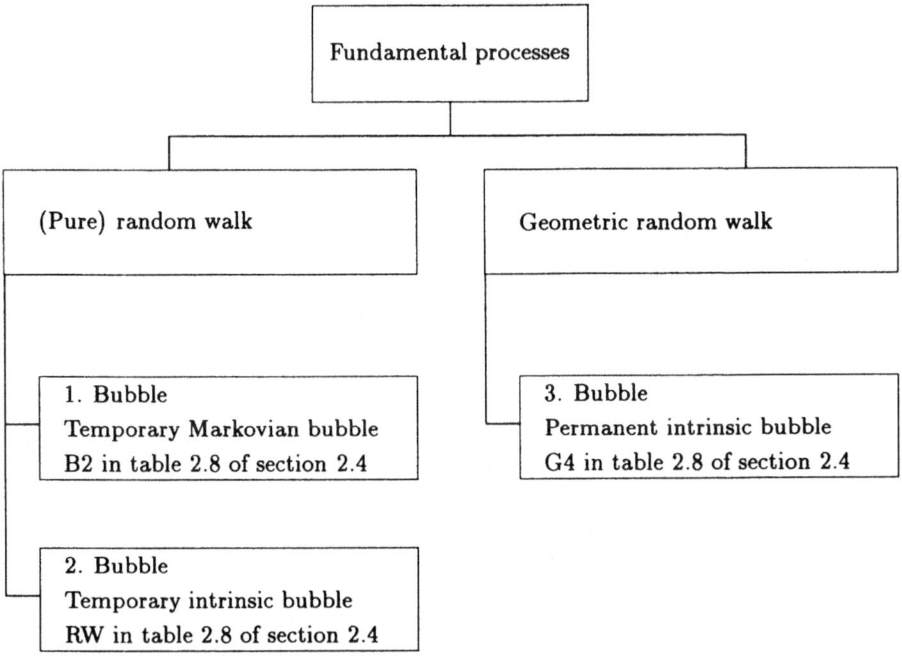

The very idea of rational bubbles capturing self–fulfilling prophecies or a self–sustaining trend behaviour may best be represented by Markovian bubbles; Markovian bubbles show ever expanding behaviour, just like fundamentally not justified expectations would predict, resulting in an extreme increase of the endogenous process.

Since a temporary occurrence of Markovian bubbles is an inherently short term effect, we use mainly quarterly data gaining more observations in comparison to annual data. Consequently, only sets 11 to 72 are considered here corresponding to class 3 to 7. We also conduct the analysis by using monthly data, the results, however, do not change significantly[13]. Based on the results in section 5.3, most fundamentals can be viewed as following a random walk except for Pi2. Consequently, set 39 is excluded from the analysis below.

Considering the Markovian bubble specification (5.1), we obtain the follow-

[13]The most interesting general result concerning monthly data is the increasing significance of the fundamental solution.

ing regression equation including two Markovian bubbles b^1 and b^2 covering the two bubble growth periods described above.

$$y_t = a + bt + c_0 x_t + c_1^t b^1 + c_2 d_t^1 + c_3^t b^2 + c_4 d_t^2 + \epsilon_t \tag{5.2}$$

$c_0 x_t$ represents the fundamental solution[14] while $c_1^t b^1$ and $c_3^t b^2$ capture the bubble solution containing time dependent coefficients. d_t^1 and d_t^2 denote dummy variables representing the duration periods mentioned above. Of course, the bubble terms themselves include dummy variables not explicitly expressed in equation (5.2), as well, due to the only temporary existence of Markovian bubbles considered here. The constant a is included because of scaling effects between endogenous and exogenous process. The trend component bt mitigates the problem of non–stationary regressors. Actually, we estimate a trend–stationary model where the estimated coefficients represent deviations from a trend. From the results of chapter 2, the economically imposed restrictions are $c_0 > 0$, $c_1 > 1$, $c_3 > 1$[15].

We estimate equation (5.2) by using non–linear OLS. The residuals are corrected for heteroskedasticity and serial correlation of unknown form. By using Newey, West's (1987) correction procedure of the variance/covariance matrix, we allow for fourth–order serial correlation and heteroskedasticity. Significant values of c_1 and c_3 will be interpreted as a hint for the possible presence of Markovian bubbles in the data.

In order to avoid strong correlation between the estimates of $c_1(c_3)$ and $b^1(b^2)$, we restrict the value of $b^1(b^2)$ to be equal to 1. This may be problematic, in particular for the estimates of $c_1(c_3)$, however, will improve the performance of the estimation procedure significantly. It may further be justified by the fact that the special deterministic intrinsic bubble process I2 in table 2.8 of section 2.4 is observationally equivalent to B2 when imposing the restriction $b = 1$.

[14]Recall the results of section 2.3 if fundamentals follow a random walk. The analysis of section 5.3 shows that some fundamental processes additionally contain a trend and a constant component. Proposition 2.10 of section 2.3 computes the corresponding fundamental solution. The resulting trend and constant components are already taken into account by including a and bt.

[15]The validity of the last two restrictions is easily seen when considering the Markovian bubble solution, i.e. $b_t = \frac{1}{\alpha^t} b$ with $\alpha = \frac{1}{1+r}$ in the case of capital markets.

The results of non–linear OLS regression are reported in appendix A.1. The estimates of the coefficients including their t–statistics are presented as well as the Durbin–Watson statistic and the coefficient of determination. Moreover, optimal starting points, denoted by 'S1' ('S2'), and stopping times, denoted by 'E1' ('E2'), for each set are given there, too. The outcome can be summarized as follows:

Class 3: Encouraging results are obtained when considering the time paths of endogenous processes contained in class 3. High values of t–statistics of both bubble term coefficients $\hat{c}_1$ and $\hat{c}_3$ indicate the high probability of the existence of Markovian bubbles in the time paths of stock indexes when national accounting data are used as fundamentals. The estimated coefficient of the fundamental process $\hat{c}_0$ is also significant for all sets, confirming the validity of the fundamental solution component. Notice that the highest significance of the fundamental component

Figure 5.4: Actual and estimated time path of set 18

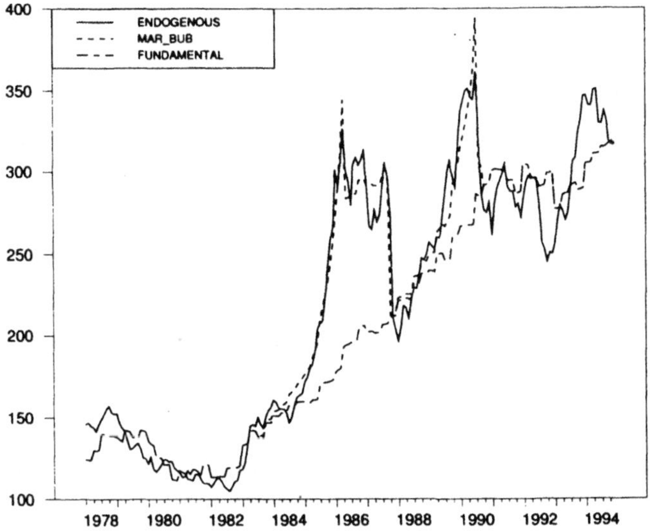

is achieved when gross capital income is used as the fundamental process. Apparently, this fundamental specification represents a good approximation for the cash flow related to stocks. Hence, both solution components explain the behaviour of the endogenous process quite well. Figure 5.4 shows the actual time path of StBA-index (indicated by 'endogenous') and the time path using the estimates of equation (5.2) where the gross capital income is used as the fundamental process (set 18) (indicated by 'mar-bub'). The time path 'fundamental' corresponds to the fundamental solution component. The similarity of the actual and the bubble time paths, especially during both bubble periods, is easily seen. These encouraging results may further be confirmed by high values of the coefficient of determination lying over 95 %. Notice further that the starting point and stopping time for sets contained in class 3 are quite similar with a starting point in the second half of 1983 until first quarter of 1984 and a stopping time in the second quarter of 1986. This is due to the fact that the underlying stock indexes and their corresponding fundamentals, i.e. national accounting data, behave very similarily. Therefore, the duration period lasts from 2:86 to 3:87. For the second bubble it can be pointed out that the starting point approximately lies in the first half of 1988 while the bubble stops growing right before the crash. Hence, the duration period is rather short or cannot be verified for the second bubble period.

Class 4: Quite similiar results are obtained for sets contained in class 4. Again, the existence of a Markovian bubble in both periods can be assured as indicated by high values of the corresponding t-statistics[16]. The validity of the fundamental solution can be confirmed since usually significant values of $\hat{c}_0$ are obtained. Therefore, the evidence of model (5.2) including fundamental and bubble solution components is high concerning most sets in class 4. The only exception is set 28 where the fundamental solution component becomes insignificant[17] and the growth period for the second bubble term proves to be rather short, casting doubt on the validity of the model (5.2) for this combination. The great explanatory power of both bubble periods may also be supported by high values of the coefficient of determination lying over 83 %. Concerning the starting point and stopping time, no clear outcome can be derived. However, it can be seen that the

[16]For set 22, 26, 27 only short growth periods are obtained. However, these growth periods are enlarging when monthly data are used.

[17]This may be due to falsly characterized fundamentals.

first bubble again ends its growth in the middle of 1986 while the starting point for the first bubble period is ambiguous. The second bubble seems to start right after the crash in 1987 and grows until mid 1990. Hence, a duration period can be verified for the former bubble while its presence is short or unlikely for the latter bubble.

Class 5: Turning to fundamentals described by proxy variables, the results change only slightly. Again, most bubble coefficients $\hat{c}_1$ and $\hat{c}_3$ are highly significant indicating the presence of Markovian bubbles in the data[18]. Again, the fundamental solution can be verified[19]. Thus, the great explanatory power of both solution components is easily seen and is further confirmed by high coefficients of determination lying between 70 % and 94 %. Figure 5.5 shows the actual and estimated time paths corresponding

Figure 5.5: Actual and estimated time path of set 31

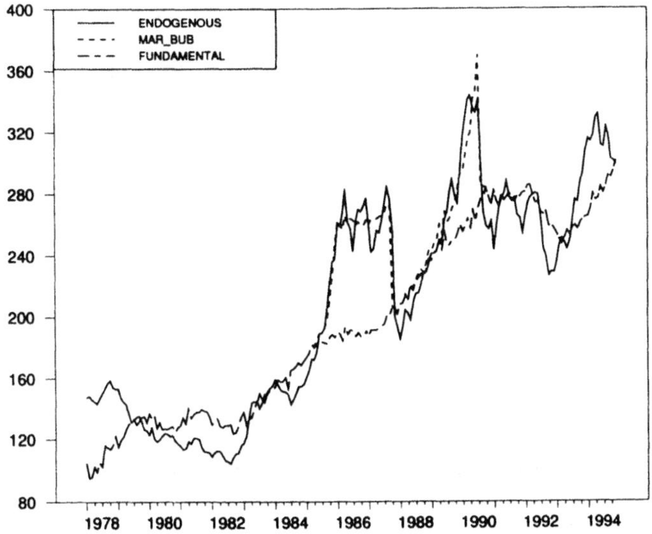

to set 31. The high coincidence of both time paths, especially during the

[18]The only exception concerns set 44 where the first bubble coefficient becomes negative.

[19]The only exceptions are set 46 – 48 where the fundamental component is estimated insignificantly. This outcome can also be interpreted as misspecified fundamentals. However, by using monthly data, the significance of the fundamental solution increases.

bubble periods, becomes obvious. The already cited results concerning the starting point and stopping time can also be acknowledged for class 5: the growth of the first bubble is likely to stop at the beginning of 1986 while a common starting point is not derivable[20]. The second bubble starts right after the crash in 1987 and increases until mid 1990. Hence, the duration period can be confirmed, at least for the first bubble, based on positive and significant values of $\hat{c}_2$. The coefficients of the second dummy are sometimes less significant which can be interpreted by the absence of a duration period or by the problematic assumption of the crash occurring for all sets at the same time. Negative values may then be seen as a hint for a prior occurrence of that crash or a possible underevaluation of the stocks after the crash.

Figure 5.6: Actual and estimated time path of set 67

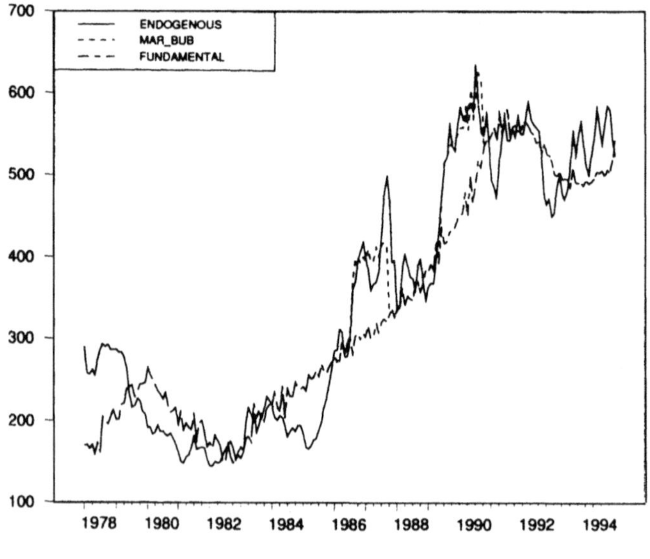

Class 6: Using single stocks as endogenous processes two different groups can be distinguished: For set 52–54 (Deutsche Babcock AG), set 61–63 (BMW AG), 64–66 (Phillip Holzmann AG), 70–72 (Mannesmann AG) the

[20]The sometimes short growth period for the first bubble increases when monthly data are used.

explanatory power of model (5.2) decreases significantly. The significance of the fundamental solution is low in all these sets, even a negative estimate for the corresponding coefficient c_0 is obtained[21]. The reason may lie in falsly characterized fundamental processes or in the presence of certain effects which are specific to these stocks, but not included by the proxy variables used in our analysis. On the other hand, many bubble terms of this group of stocks are significantly estimated[22]. However, we doubt the relevance of model (5.2) for these stocks because fundamentals are likely to be misspecified. Since the concept of rational bubbles depends decisively on the correct characterization of the fundamentals, the results concerning the bubble terms may be misleading. In contrast, the outcome concerning the remaining stocks, i.e. set 55–57 (BASF AG), set 58–60 (Bayer AG) and set 67–69 (Karstadt AG) is more affirmative. The significance of most fundamental solution components can be verified[23]. The existence of a Markovian bubble in both periods can be confirmed, indicated by highly significant estimates for c_1 and c_3[24]. Consequently, some evidence for the relevance of model (5.2) is provided for these stocks, due to significant estimates of both solution components. Figure 5.6 shows the actual and estimated time path of set 67. Both bubble periods are well captured by the estimated time paths[25]. Moreover, R^2 varies between 18 % and 88 % for the first group and between 91 % and 93 % for the second group, confirming the high explanatory power of Markovian bubbles for the latter. Finally, there is little support for common timing concerning the different bubble periods. Both starting point and stopping time of both bubble periods vary extensively within the considered interval. However, for the

---

[21]Notice that the significance of $\hat{c}_0$ can be increased for set 64–66 when monthly data are used.

[22]Exceptions are set 53 where both bubble coefficients are negative and set 64–66 where the first bubble coefficient is insignificant. When monthly data are used, the number of negative or insignificant bubble terms increases.

[23]The partly low significance of set 55 – 60 is increased when monthly data are used.

[24]The only exception is the second bubble of set 68 where a negative coefficient is obtained. Notice further that the short growth period for set 55–60 is enlarged extensively when monthly data are used.

[25]Notice that the crash of the first bubble appears to occur later than indicated by our analysis.

latter group of stocks the timing results of the previous classes can often be confirmed.

Concerning the results summarized above, the following *remarks* should be mentioned:

- Taking a look at appendix A.1, the extreme autocorrelation, indicated by low values of the DW–statistic, becomes apparent. The procedure applied in the analysis above for correcting for autocorrelation is inferior to other procedures if the precise degree of autocorrelation is known. Since this is not the case here, the application of Newey, West's (1987) correction method is appropriate. Moreover, the eventual biasedness of the estimates is supposed to have only a small impact on the significance of the bubble estimates since the t–statistics generally show extremely high values.

- Until now, nothing has been said about causality. As mentioned at the beginning of chapter 2, equation (2.4), considered there, introduces a causality in the sense that the endogenous process is determined by the exogenous process and not the other way around. For testing the converse assumption, a battery of Granger–Causality tests has been conducted. The results are not reported here explicitly. However, they can be summarized by stating that, for most of the sets, the endogenous process has no explanatory power for the fundamental process[26].

Summarizing, the first bubble test has provided some evidence for the existence of Markovian bubbles in the data when indexes or sector indexes are used as endogenous processes and national accounting data or proxy variables are used as fundamentals. Markovian bubbles have at least the potential to explain the strong deviation of the endogenous process from its fundamentals within both bubble periods 1980–1987 and 1987–1990. For

[26] We conduct traditional Granger–Causality tests by applying the standard final prediction error–argument in order to determine the number of lags included, first, for the fundamental process alone, then for the inclusion of lagged values of the endogenous process. Finally, an F–test concerning the inclusion of all lagged values of the endogenous variable should be rejected if the direction of causality is correctly assumed.

single stocks, however, the evidence is mixed and the existence of rational bubbles is difficult to assure. Stock-specific effects are likely to exist and to interfere with the impact of rational (Markovian) bubbles.

5.2.2 Temporary intrinsic bubbles

The second bubble specification to be tested for is a temporary intrinsic bubble. Based on the results of table 2.8 of section 2.4, the intrinsic bubble process corresponding to fundamentals which follow a random walk, can be expressed as

$$b_t = \frac{1}{\alpha^t} x_t \tag{5.3}$$

In analogy to the Markovian bubble considered previously, the bubble specification (5.3), due to its exponential term included, may equivalently explain the extreme deviation of endogenous and exogenous process. Moreover, the specification as an intrinsic bubble, depending on the fundamental process x_t, additionally covers the idea of overreaction to fundamental news if $\alpha < 1$.

Based on the intrinsic bubble specification (5.3), the regression equation differs only slightly from the bubble test conducted above and can be stated as

$$y_t = a + bt + c_0 x_t + c_1^t x_t + c_2 d_t^1 + c_3^t x_t + c_4 d_t^2 + \epsilon_t \tag{5.4}$$

The only difference is that, instead of Markovian bubble terms b^1 and b^2, the fundamental process x_t is introduced as an additional temporary intrinsic bubble component including time driven coefficients. However, instead of equation (5.4), the price–dividend ratio is tested for the presence of temporary intrinsic bubbles in order to avoid the problem of multicollinearity. Moreover, it turns out that this transformation improves the estimation results significantly. By dividing equation (5.4) through x_t, the following regression equation is obtained.

$$\frac{y_t}{x_t} = a\frac{1}{x_t} + b\frac{t}{x_t} + c_0 + c_1^t + c_2 d_t^1 + c_3^t + c_4 d_t^2 + v_t \tag{5.5}$$

where $v_t = \frac{\epsilon_t}{x_t}$. All other components are already introduced above. From the results of chapter 2, economically imposed restrictions are $c_0 > 0$,

$c_1 > 1$, $c_3 > 1$[27].

Again, non–linear OLS is used including Newey, West's (1987) correction procedure as presented above. Moreover, the same data are used here, i.e. set 11–72. While the following discussion mainly relates to results, when using quarterly data, the analysis is also carried out for monthly data, however, only little changes in estimation results are obtained. The outcome of non–linear OLS regression, i.e. estimates including their t–statistics, is reported in appendix A.2. Additionally, besides the DW–statistic and the coefficient of determination, optimal starting points 'S1' ('S2') and stopping times 'E1' ('E2') are tabulated there. The results can be interpreted as follows:

Figure 5.7: Actual and estimated time path of set 18

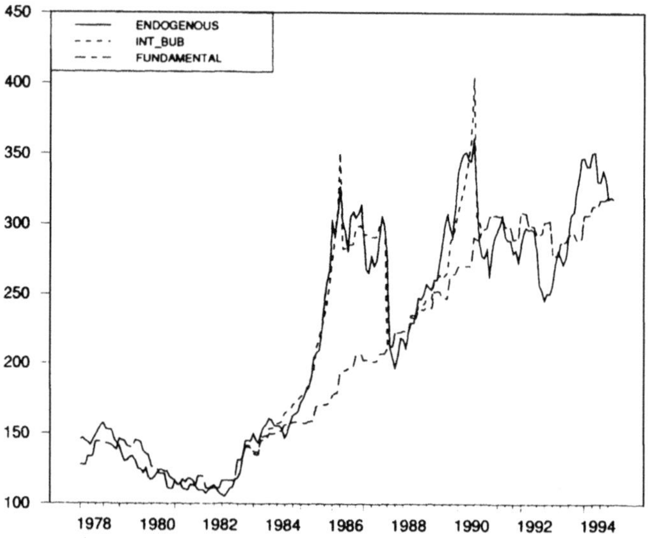

Class 3: Interestingly, the affirmative results of the former Markovian bubble specification can mostly be replicated when model (5.5) is estimated. Bubble coefficients are highly significant for all sets contained in

[27]The validity of the last two restrictions is easily seen when considering the intrinsic bubble solution for fundamentals following a random walk, i.e. $b_t = \frac{1}{\alpha^t} x_t$ with $\alpha = \frac{1}{1+r}$ in the case of capital markets.

class 3 as well as the fundamental solution[28]. The similarity of the regression results between both bubble specifications can also be confirmed by looking at the timing of the bubble periods considered. For almost all sets, equal starting points and stopping times are obtained. It becomes obvious that both bubble specifications, Markovian and intrinsic, are equally capable of explaining the strong deviation of the endogenous variable from its fundamentals during both bubble periods considered. Only the coefficient of determination decreases a little lying between 89 – 94 %. Figure 5.7 shows the actual and estimated time paths of set 18 where 'int–bub' denotes the estimated time path including the intrinsic bubble specifications. The explanatory power of model (5.5) can easily be seen. Notice again that comparing figure 5.7 and figure 5.4 only slight differences become apparent.

Class 4: Concerning sets contained in class 4, the evidence of temporary intrinsic bubbles is again in most cases affirmative. All bubble components are estimated significantly and satisfy the restrictions imposed on economic grounds. The only exception concerns set 28 where the growth period takes only one quarter of the year[29]. Also the fundamental solution components exhibit reasonable and significant estimates. Principally the same results, as found in the first bubble specification, are obtained concerning the timing with a difference of only one or two quarters of the year. However, the coefficient of determination varies between 73 – 92 % indicating a lower fit of the model in comparison to the Markovian bubble model.

Class 5: When using proxy variables as fundamentals, the similarity between both bubble specifications already mentioned remains valid. The existence of all bubble components can be confirmed, due to positive, and significant, estimates of c_1 and c_3 being greater than one[30]. All funda-

[28]Notice that again the best result concerning the fundamental solution is obtained when gross capital income is used as fundamental process.

[29]Notice, however, that the growth period enlarges when monthly data are used.

[30]The only exception is again set 44 where the first bubble exhibits a negative coefficient, violating the economic restriction of being greater than one.

Figure 5.8: Actual and estimated time path of set 31

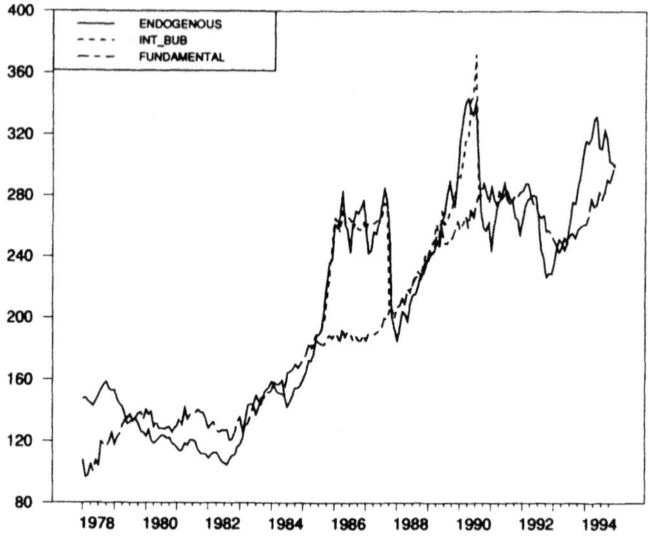

mental solution components can be verified[31]. The timing of the bubble periods is similar to the Markovian bubble tested for previously, however, the fit measured by R^2 is lower for the instrinsic bubble specification lying between 41–90 %. Figure 5.8 shows the actual and estimated time path of set 31. The similarity of both time paths is obvious. Notice that figure 5.8 resembles figure 5.5 heavily.

Class 6: When single stocks are considered as endogenous processes, the ambiguous results as in the case of Markovian bubbles can also be emphasized for the instrinsic bubble case. The low significance of the fundamental solution, combined with mostly insignificantly or negatively estimated bubble terms, casts doubt on the relevance of model (5.5) for set 52–54 (Babcock AG), set 61–63 (BMW AG), set 64–66 (Phillip Holzmann AG) and set 70–72 (Mannesmann AG). However, notice that for the last three stocks, at least the second bubble terms exhibits its economically expected behaviour. On the other side, concerning the remaining stocks, i.e. set 55–57 (BASF AG), set 58–60 (Bayer AG) and set 67–69 (Karstadt

[31] Again, for set 46–48 c_0 is estimated at a low level of significance.

AG), the evidence is more affirmative. The relevance of all solution components, fundamental as well as both bubble terms, can be verified since significant estimates satisfying the economic restrictions are generally obtained[32]. The timing is also similar to the former bubble specification, at least for the second group of stocks, while the coefficient of determination decreases differing from 21 % to 84 % for the first group and from 85 % to 90 % for the second. Figure 5.9 shows the actual and estimated time paths of set 67. Comparing figure 5.9 with figure 5.6, again, the similarity between both bubble specification is obvious.

Figure 5.9: Actual and estimated time path of set 67

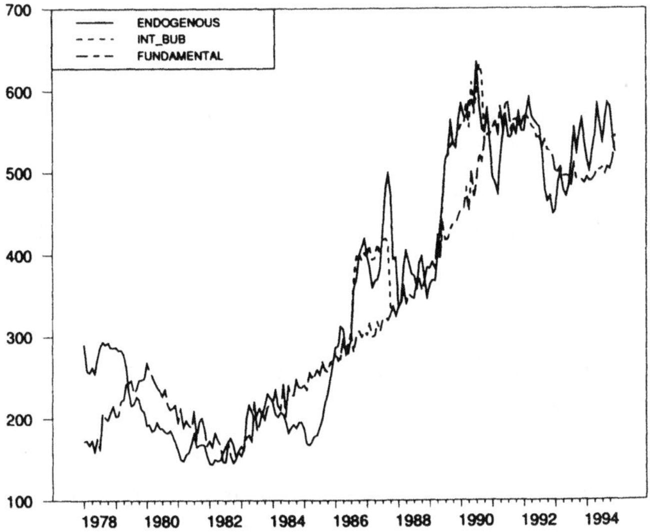

Summarizing, the extreme deviation of the endogenous process from its fundamentals during both bubble periods can equivalently be explained by the temporary existence of the special intrinsic bubble (5.3), especially when indexes are used as endogenous variable. The similarity to the former bubble test is due to the fact that the intrinsic bubble specification assembles the Markovian bubble specification in the sense that it is also able to

[32]The only exception is set 68 where a negative coefficient is obtained for the second bubble period. Notice further that the short growth period of set 55–60 can be enlarged when monthly data are used.

explain the trend behaviour captured by the exponential terms included in both specifications. However, notice that generally the explanatory power of the instrinsic bubble specification is lower than in the case of a Markovian bubble. This may be interpreted as a hint that the extrapolative trend effect dominates the German Stock Market, as compared to the possibility of a prevailing overreaction effect captured by the intrinsic bubble specification.

Concluding, comparing both temporary bubble specifications, the Markovian bubble representing the trend behaviour exclusively, is more likely to explain the movement of stock prices correctly than the intrinsic bubble specification when considering both a trend effect and additionally an overreaction effect.

5.2.3 Permanent intrinsic bubble

The third bubble specification to be tested for is a permanent geometric intrinsic bubble II process (G4 of table 2.8 of section 2.4) and can be described as

$$b_t = x_t^\lambda \tag{5.6}$$

The application of this bubble specification can economically be justified again because of its potential for explaining the observed excessive deviation between the endogenous and exogenous process. Moreover, it covers the idea of overreaction to fundamental news if λ turns out to be positive and greater than 1. Notice that due to the restrictions imposed, in order to derive this bubble type[33], no time driven coefficient is present here which enables the examination of a permanent bubble unlike both preceding bubble specifications. Recall further that fundamentals need to follow a geometric random walk in this case.

Froot, Obstfeld (1991b) first test for the existence of this bubble type by using US–stock market data. Bruns (1994) applies this testing approach for examining the existence of intrinsic bubbles on the German Stock Market. The analysis conducted here differs from the analysis conducted by Bruns (1994) in the following aspects:

[33] See the derivation of intrinsic bubble in section 2.4.

- Instead of merely using a stock index as the endogenous process, we consider also sector indexes and single stocks in our analysis. Moreover, instead of merely using dividends as fundamentals, we extend the analysis by additionally considering national accounting data and proxy variables as fundamentals.
- Besides traditional t–tests, on which Bruns (1994) analysis is mainly based, a likelihood–ratio test is conducted, as suggested by Froot, Obstfeld (1991b), in order to gain evidence for the explanatory power of the bubble component.
- No preliminary investigation into whether fundamentals can be described by a geometric random walk, was conducted by Bruns (1994). We provide this preliminary analysis in section 5.3.

Based on the particular solution for fundamentals following a geometric random walk in proposition 2.10 of section 2.3 and the bubble specification (5.6), the following non–linear relationship between endogenous and fundamental processes is estimated by non–linear OLS

$$y_t = c_0 x_t + c_1 x_t^{\lambda} + \epsilon_t \tag{5.7}$$

$c_0 x_t$ denotes the fundamental solution and $c_1 x_t^{\lambda}$ the bubble solution as derived previously[34]. The economically imposed restrictions are $c_0 > 0$, $c_1 > 0$, $\lambda > 1$.

When estimating equation (5.7), the residuals are corrected for heteroskedasticity and serial correlation of unknown form again by using Newey, West's (1987) correction procedure of the variance/covariance matrix, allowing for fourth–order serial correlation and heteroskedasticity. Like Froot, Obstfeld (1991b) common t–statistics of estimated parameters are calculated for statistical inference. Froot, Obstfeld (1991b) have shown that under certain conditions[35] the t–statistic of $\hat{c}_1$ is nearly distributed standard normally. Additionally, a likelihood ratio test is carried out.

[34]Additionally, in order to correct for possible multicollinearity, the ratio of endogenous to fundamental processes is estimated by non–linear OLS, as well. However, it turned out that the estimation results do not differ much between both regression types. Hence, we will only report the results of non–linear OLS–regression of equation (5.7).

[35]See section 4.2.2.

Testing for the permanent intrinsic bubble as considered here is only justified if fundamentals follow a geometric random walk. The presumption of non–stationary log–fundamentals is tested for in section 5.3. Results obtained there provide the basis of the analysis of permanent intrinsic bubbles here. By summarizing the outcome of integration tests, table 5.5 of section 5.3 shows the sets reasonably excluded from testing for a permanent intrinsic bubble, i.e. sets 4, 5, 6, 7, 10, 39.

Recall that in the case of capital markets with $\alpha = \exp(-r)$, c_0 can theoretically be described as $(\exp(r) - \exp(\mu + \sigma^2/2))^{-1}$, according to proposition 2.10 of section 2.3, and λ relates to the positive root of the characteristic equation $\lambda\mu + \frac{1}{2}\lambda^2\sigma_\nu^2 - r = 0$. μ and σ^2 are the distributional parameters of the underlying fundamental process following a geometric random walk and r stands for an average one–period real rate of return. In our analysis we compute the theoretically warranted values of c_0 and λ and compare them with those obtained by regressing equation (5.7). Additionally, in order to take account for the correlation between the bubble parameters c_1 and λ, we conduct a restricted regression by using the theoretically derived value for λ when estimating equation (5.7). Notice, however, that both computing the theoretically expected values and conducting a restricted regression are only possible for sets relating prices and dividends since in this case dividends represent an actual cash flow connected with the stock considered. Thus, these computations are only undertaken for set 1 to 10. The results are tabulated in appendix A.3 and A.4 and can be summarized as follows:

Class 1: For class 1 it turns out that the parameter $\hat{c}_0$ is highly significant and close to its theoretically expected value. This may be interpreted as strong evidence for the validity of the fundamental solution. However, concerning the bubble solution no significant impact can be pointed out, indicating that a permanent intrinsic bubble is unlikely to be present in the data, although the deviation of $\hat{\lambda}$ from its theoretical value appears to be unexpectedly small (see, in particular, set 3). Applying restricted regression, the results become more interesting. While the fundamental part remains almost unchanged, we obtain reasonable, i.e. positive values for the parameter $\hat{c}_1$ when using the theoretical value of λ in the regres-

Figure 5.10: Actual and estimated time path of set 3 – restricted case

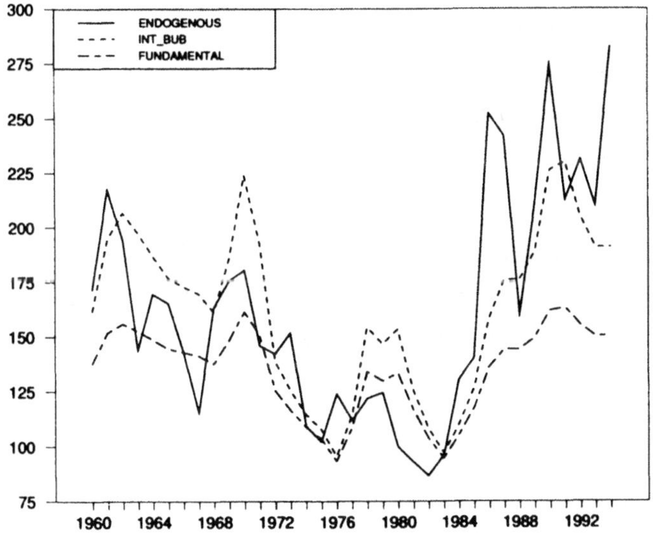

sion. This may indicate the presence of bubbles, however, at a low level of significance. Figure 5.10 shows the actual and estimated time paths of set 3 in the restricted case. It comes out that at least a part of the excessive time path behaviour of stock market indexes is captured by the intrinsic bubble specification, in particular during 1985–1991.

Class 2: The estimation results of class 2 are disappointing[36]. Concerning unrestricted regression, the fundamental component indicated by the parameter $\hat{c}_0$ becomes insignificant in most cases, even negative extremely violating the theoretically warranted parameter restrictions[37]. While the

[36]Notice that, although sets 4, 5, 6, 7, 10 are excluded from our regression as a result of integration tests, we report the results of all sets contained in class 2 since these sets appear to behave rather similarly no matter whether the fundamentals can be described as a geometric random walk or not.

[37]It seems that dividend processes have only a small or economically unreasonable impact on the corresponding stock prices when, in addition, permanent intrinsic bubbles are considered, at least in the unrestricted case. However, we must bear in mind that most dividend processes cannot be described as a geometric random walk such that this failure may have caused the bad results.

positive estimates of $\hat{c}_1$ prove to be insignificant in most cases, λ turns out to be both significant and insignificant. Additionally, no coincidence of estimated and theoretical values is obtained. The results hardly changes if regression is restricted. Fundamental components become highly significant and positive, however, $\hat{c}_1$ while significant in most cases, takes negative values violating the economically imposed restriction $c_1 > 0$.

Class 3–5: The results of classes 3 to 5 turned out to be very similar and are, therefore, treated here together. First, the high significance of the fundamental component can be emphasized for most sets. All proxy variables appear to have a strong impact on the endogenous process. The bubble parameter $\hat{c}_1$, however, is highly insignificant and takes economically unreasonable negative values. Although λ is for the most part significantly estimated, a negative parameter λ and, hence, a permanent intrinsic bubble is of minor relevance in these classes. Figure 5.11 shows the actual

Figure 5.11: Actual and estimated time path of set 31

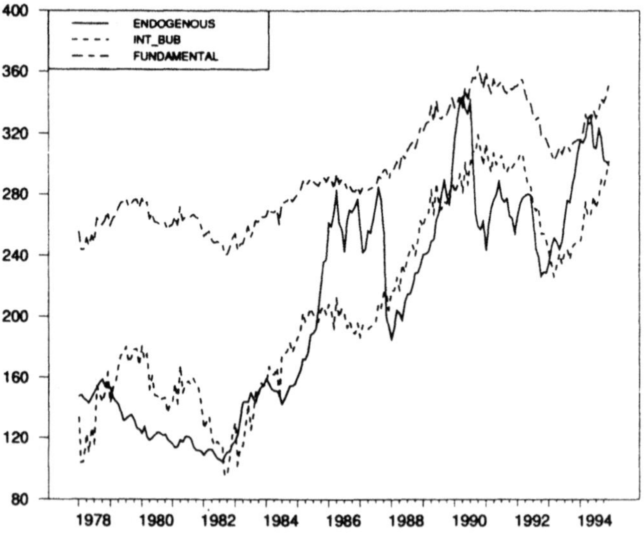

and estimated time path of set 31. The large deviation of the fundamental solution, lying extremely above actual market prices, from the bubble solution, especially during 1984–1987 and 1988–1990, becomes apparent.

Moreover, the model possesses only low explanatory power, in particular during the bubble periods.

Class 6: By using single stocks rather than indexes, the quality of the results once again decreases considerably. In class 6 the validity of the fundamental component can only be verified at low levels, in some cases the estimates take even negative values. This replicates the disappointing results of class 2. The estimates of the lasting parameters remain unsatisfactory, as well, i.e. negative for c_1 and λ, the former insignificant, the latter mostly significant. Again the presence of permanent intrinsic bubbles could not be shown.

As already mentioned, the problem of estimating equation (5.7) is the high correlation between c_1 and λ. In order to take this interdependence into account, an additional likelihood–ratio test is proposed by Froot, Obstfeld (1991b). Concerning our analysis it turns out that in most cases the bubble component appears to have indeed a significant impact in determining the behaviour of the endogenous process[38]. However, the outcome of the LR–test does not change our interpretation of the results against the existence of permanent intrinsic bubbles. The negative influence of most bubble components, contradicting the theoretically expected effect, can be seen as the most striking counterargument against the presence of permanent intrinsic bubbles in the data, at least if intrinsic bubbles aim to capture the idea of overreaction. This interpretation can graphically be supported by figure 5.11 where the fundamental solution exceeds, to the extreme, the actual time path.

Summarizing, the existence of permanent intrinsic bubbles as described in equation (5.7) is extremely unlikely[39]. The results of non–linear OLS estimation lead to the conclusion that for most sets the permanent presence of this special intrinsic bubble[40] cannot be proven, especially when considering single stocks. The low explanatory power of the fundamental

[38]This is the main difference to the analysis of Bruns (1994).

[39]Bruns (1994) states a similar outcome.

[40]We also tested for a temporary existence of this bubble specification, however, the results remain unchanged.

solution, combined with the economically unreasonable impact of the bubble component, disputes the appropriateness of model (5.7) for explaining the movement of stock prices. The only encouraging result concerns set 1 to 3 where a restricted regression provides at least some evidence in favour of the existence of permanent intrinsic bubbles. On the other side, the fundamental solution could be verified in most sets at high levels of significance.

All three preceding test procedures rely on certain assumptions concerning the fundamental processes involved. The subsequent digression presents common techniques for unit root tests, applies these techniques to the processes considered and discusses their results as a basis for the analysis conducted above.

5.3 A digression: Testing for unit roots

Various strategies in testing for unit roots in univariate time series[41] exist in economic literature[42]. In this digression we provide a comprehensive decision procedure which enables us to distinguish between different model specifications[43]. In general, in order to test for unit roots, three different auxiliary regressions may theoretically be considered. The first auxiliary regression which we will refer to as model I includes both a constant and a trend component representing the most unrestricted model specification.

$$z_t = \hat{\rho} z_{t-1} + \hat{\mu} + \hat{\beta} t + v_t \tag{5.8}$$

---

[41]There exist other procedures for testing for unit roots in multivariate time series. See, for instance, Johannsen (1988, 1992).

[42]For a general description of concepts of integration and cointegration see, for instance, Banerjee, Hendry (1992) and Dolado et al. (1991).

[43]Notice that we do not provide a strategy that takes into account techiques of traditional time series analysis. Therefore, a more general procedure may exist which explains the nature of the time path considered more precisely. Here, we are only concerned with the question whether a process can be regarded as stationary or as non–stationary.

z_t stands for an arbitrary stochastic process, here for both an endogenous and exogenous process, μ denotes a constant and t a time trend including the corresponding parameter β. The parameter ρ represents the decisive coefficient. For $\rho = 1$, z_t follows a random walk, for $\rho < 1$, an AR(1)–process is appropriate. The second auxiliary regression which we will refer to as model II considers only a constant and can, therefore, be described as

$$z_t = \tilde{\rho} z_{t-1} + \tilde{\mu} + v_t \tag{5.9}$$

while model III takes only the lagged value of the process under consideration into account.

$$z_t = \breve{\rho} z_{t-1} + v_t \tag{5.10}$$

The signs above the parameters indicate estimated coefficients which form the basis of the following testing procedure. Consequently, v_t represents the regression residuals related to the corresponding model specification. Model I to III is estimated by OLS.

For statistical inference in the context of unit root tests, the auxiliary regressions (5.8) to (5.10) are generally confronted with the null hypothesis that process z_t follows a (pure) random walk, i.e.

$$H_0 : z_t = z_{t-1} + \epsilon_t \qquad H_1 : z_t = \rho z_{t-1}[+\mu + \beta t] + \epsilon_t$$

where ϵ_t stands for an iid–error term. For gaining evidence of this hypothesis, we calculate two groups of test statistics based on the estimates of the decisive parameter ρ and other parameters β and μ, both taking into account possible heteroscedasticity and autocorrelation of unknown form in the residuals.

- First, we compute an augmented Dickey, Fuller (ADF)–statistic[44] of the estimates $\hat{\rho}$, $\tilde{\rho}$, $\breve{\rho}$ by incorporating 4 lags of differenced values of the process under consideration in model I to III. These statistics will be denoted as $t_{\hat{\rho}}$, $t_{\tilde{\rho}}$, $t_{\breve{\rho}}$. It is well known that the various ADF–statistics have a degenerated distribution tabulated in Fuller (1976) if the null hypothesis is a pure random walk. Additionally, the ADF–statistics of the other coefficients denoted by $t_{\hat{\beta}}$ $t_{\hat{\mu}}$, $t_{\tilde{\mu}}$ are calculated.

[44]See Dickey, Fuller (1979, 1981).

Under the null hypothesis of a pure random walk, their distribution is degenerated, as well, and tabulated in Dickey, Fuller (1981).

- Second, other statistics of parameter ρ correct for possible heteroscedasticity by using a Newey, West's (1987) correction of the variance/covariance matrix, as proposed by Phillips (1987) and Phillips, Perron (1988). These PP–statistics will be denoted by $Z(t_{\hat{\rho}})$, $Z(t_{\tilde{\rho}})$, $Z(t_{\breve{\rho}})$. They have the same distributional properties as the ADF–statistics. Additionally, the PP–statistics of the other coefficients denoted by $Z(t_{\hat{\beta}})$ $Z(t_{\hat{\mu}})$, $Z(t_{\tilde{\mu}})$ are calculated. Their critical values are again tabulated in literature cited above.

Notice that all statistics considered until now have a degenerated distribution which comes from the fact that the null hypothesis relies on the assumption that the true data generating process is a random walk. But their distributional properties change considerably if we assume a random walk with drift as null hypothesis in the sense that their limiting distribution becomes normal.

$$H_0 : z_t = \mu + z_{t-1} + \epsilon_t \qquad H_1 : z_t = \rho z_{t-1} + \mu[+\beta t] + \epsilon_t$$

However, as was first pointed out by Hylleberg, Mizon (1989) and Haldrup, Hylleberg (1989), if the sample size is small the distributional characteristics deviate from the normal[45]. The small sample distribution of the various t–statistics has been derived by Hylleberg, Mizon (1989, 1995) and Haldrup, Hylleberg (1989). In order to gain evidence for a random walk with drift hypothesis, we run the following preliminary OLS–regression for the first difference of each process Δz_t[46], i.e.

$$\Delta z_t = \hat{m} + v_t \tag{5.11}$$

where $\hat{m}$ stands for the estimate of the drift component and v_t denotes the residuals of equation (5.11). The t–statistic of $\hat{m}$ is standard normally

[45]Using model II, for instance, the distribution of the estimated parameter ρ varies between the values tabulated in Fuller (1976) and the standard normal distribution approximating the latter if sample size increases.

[46]This procedure was first proposed by Haldrup, Hylleberg (1989). See also Kohn (1993) for an application of integration tests including this preliminary regression in the context of exchange rates.

distributed under the null. If the constant is significant, we use the normalized estimate of the constant $\hat{m}/\hat{\sigma}_v$ as a proxy for the μ value in the tables of Haldrup, Hylleberg (1989) and Hylleberg, Mizon (1989).

Figure 5.12: Preliminary Regression

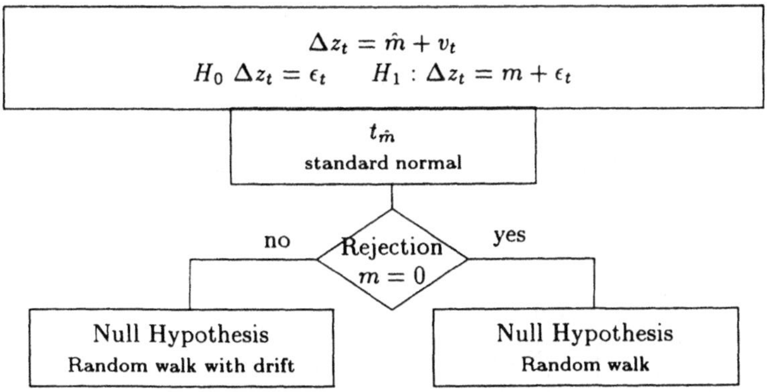

For greater clarity, appendix B.1 provides a full characterization of each single t–statistic and its small sample and limiting distribution. Critical values of the corresponding t–statistics under various null hypothesis are summarized in appendix B.2.

Based on the different auxiliary regression specifications (5.8) to (5.10), we can apply the following decision procedure shown in figures 5.12 to 5.15. First, the preliminary regression (5.11) is run for every variable considered (see figure 5.12). If the drift term is significant, the small sample critical values derived by Haldrup, Hylleberg (1989) and Hylleberg, Mizon (1989) are applied. If the drift term is insignificant, the critical values derived by Fuller (1976) and Dickey, Fuller (1981) are used instead. In the following flow charts the changes in testing for a random walk with drift rather than for a (pure) random walk, are encapsulated in parathesis '{}'.

Then, we continue by estimating the most unrestricted model I by OLS (figure 5.13). If the null hypothesis of a unit root can be rejected[47], it is tested for the significance of the trend in equation (5.8). Note that in

[47]Notice that the distribution of the statistic $t_{\hat{\rho}}$ is independent of the null hypothesis being a pure random walk or a random walk with drift.

Figure 5.13: Testing in model I

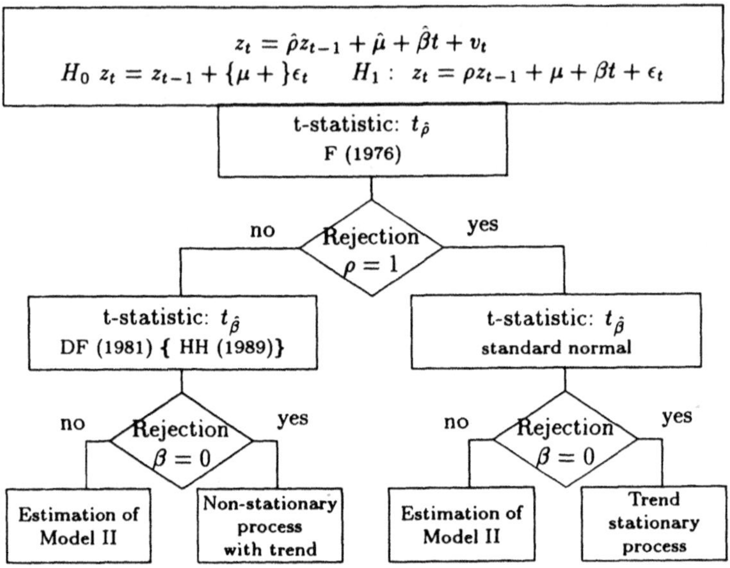

Remarks: F = Fuller; DF = Dickey, Fuller; HH = Haldrup, Hylleberg
Modifications, due to a null containing a drift term, are labeled by '{}'.

this case the t–statistic of $\hat{\beta}$ is standard normally distributed. If the trend coefficient appears to be significant, the stochastic process considered may be viewed as trend stationary. Otherwise, model I has to be re–estimated without the trend component, leading up to model II[48].

However, if the null hypothesis of a unit root in model I cannot be rejected, we again test for the significance of the trend term. If the null hypothesis contains a constant (significant $\hat{m}$ in equation (5.11)), $t_{\hat{\beta}}$ is distributed as tabulated in Haldrup, Hylleberg (1989). If the constant in equation (5.11) is not significant, $t_{\hat{\beta}}$ is distributed as tabulated in Dickey, Fuller (1981). According to the corresponding critical values, we stop if the trend coefficient is significant, claiming that the stochastic process is considered

[48]Notice that when applying this procedure of sequential or nested hypothesis testing, the overall level of significance differs from the level of significance on each single stage. For instance, if the test is conducted at a 5 % level on each stage, the overall level is (1-5%)(1-5%). We can mitigate this problem by using a lower significance level.

as non-stationary with trend (and eventually a constant). Otherwise, we go further by estimating model II.

Figure 5.14: Testing in model II

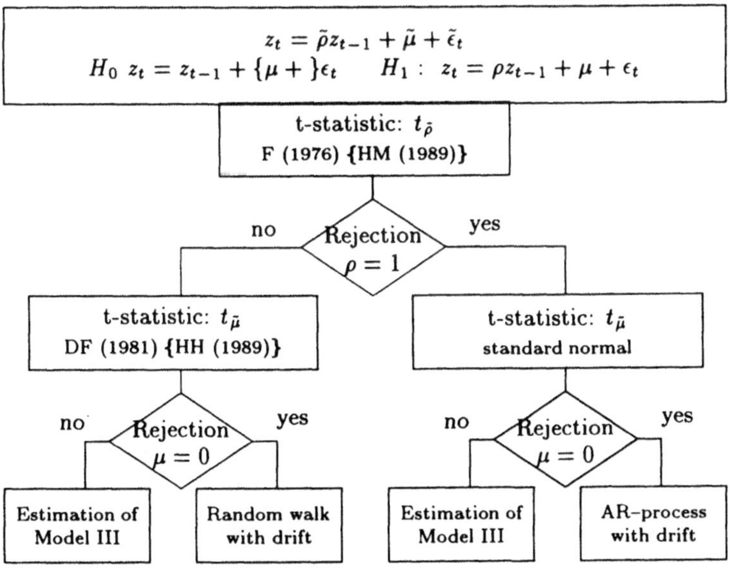

Remarks: F = Fuller; DF = Dickey, Fuller; HM = Hylleberg, Mizon; HH = Haldrup, Hylleberg
Modifications, due to a null containing a drift term, are labeled by {}.

An analogous procedure can be applied to the set of estimates concerning model II which omits the trend component in the auxiliary regression. The precise flow chart is shown in table 5.14. The decision procedure concerning the third auxiliary regression specification is trivial and displayed in table 5.15 only for completeness.

Before reporting the results and discussing their implications, a note concerning greater differences between the proposed statistics (ADF– and PP–statistics), sometimes observed in the results, appears to be necessary. These discrepancies may mainly be due to two reasons:

- As already pointed out by Phillips, Perron (1989), the PP–statistic have low power in the case of negatively correlated variables. The application of ADF–statistics are recommended here instead.

Figure 5.15: Testing in model III

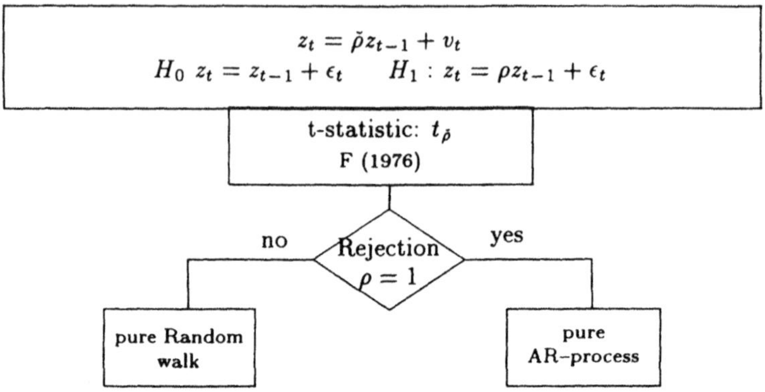

Remarks: F = Fuller

- On the other hand, ADF–statistics may include an incorrect number of lags of the variable under consideration. This may lead to an inappropriate correction for autocorrelation. Here, the application of PP–statistics are recommended instead.

If greater differences between both statistics occur, further analysis, taking into account the above mentioned caveats, must be conducted. However, in most cases it will turn out that both statistics take very similar values. In the following we apply this procedure to all processes involved in our analysis.

Results of integration tests: The results of the *preliminary regression* (5.11) are tabulated in appendix A.5.1 and A.5.2. Equation (5.11) is estimated by OLS including varying numbers of lagged differences, of the variable under consideration in order to correct for possible autocorrelation. Appendix A.5.1 and A.5.2 report the estimated parameter $\hat{m}$ and its t–statistic in parenthesis beneath the estimate. The number of lags, added for correction purposes, and the corresponding proxy μ for using the tables derived by Haldrup, Hylleberg (1989) and Hylleberg, Mizon (1989) are also included. Significance is measured at a 5 % level. Based on these results, it turns out that, for the majority of the processes considered, the null hypothesis of no random walk with drift cannot be rejected, i.e. the critical values derived by Fuller (1976) and Dickey, Fuller (1981) are applicable. In particular, from the tables we can draw the following conclusions:

App. A.5.1: Concerning *endogenous processes* only the stock processes St2, St3 and St4 contain significant drift components, implying approximated values of μ around 0.01 which does not lead to any significant change in critical values. Consequently, concerning the entire set of endogenous processes, the Dickey, Fuller set of critical values should be applied.

App. A.5.2: Concerning *exogenous processes* the following results can be pointed out: The whole set of national accounting data (Na1, Na2, Na3, Na4, Na5, Na6) can be viewed as possibly following a random walk with drift[49]. Since the proxy μ differs significantly from values around 0.01, we are forced to apply critical values calculated by Haldrup, Hylleberg (1989) and Hylleberg, Mizon (1989). Moreover, Div2, Pi2, Io2, It1, It2, It7 contain significant estimates of m partly leading to high values of the proxy μ subsequently used in the tables cited above. Summarizing, for Div2, Na1, Na2, Na3, Na4, Na5, Na6, Pi2, Io2, It1, It2, It7 we use tables proposed by Haldrup, Hylleberg (1989) and Hylleberg, Mizon (1989), for the remaining processes traditional critical values of Dickey, Fuller (1979, 1981) apply.

After conducting the preliminary regression, all processes considered are tested for the presence of a unit root. The results of *integration tests* are tabulated in appendix A.5.3. We report both the estimates of model I to III and its t–statistics (ADF– and PP–statistics) beneath the estimates in parenthesis. Significance is measured at a 10 % level.

App. A.5.3: Concerning *endogenous processes* the following conclusions can be drawn: Indexes In1, In2, In3 may best be explained by incorporating

[49]However, if we run the same regression (5.11) including a time trend the constant term becomes insignificant. This may reflect the fact that national accounting data may be seen as trended data rather than as a random walk with drift. The drift term in equation (5.11) then incorporates most of the effects of the trendiness of the data.

a trend component and containing a unit root. Notice, however, that the evidence for In1 is mixed. Sector–indexes Se1, Se4, Se5 prove to be non–stationary with trend component while Se2, Se3, Se7, Se8, Se10 seem to follow a pure random walk. Se6 can be viewed as non–stationary containing a constant[50]. Finally, the evidence for Se9 is mixed, due to contradicting ADF– and PP–statistics. However, further analysis shows that the PP–statistic appears to be more appropriate implying the non–stationarity with trend component of Se9. Stocks St2, St3, St4, St6, St7 are likely to follow a random walk with trend while St1 seems to be best explained by an autoregressive process containing a constant. Finally, the evidence concerning St5 is mixed. However, further analysis shows that St5 may best be described as following a pure random walk plus a trend component. Summarizing, we only regard St1 as stationary. All other endogenous processes are describable as random walks.

Considering *exogenous processes* we state the following conclusions, taking into account the results derived in the preceding subsection concerning the random walk with drift hypothesis: Except for Div1 which indeed seems to follow a (pure) random walk the remaining dividend processes can be considered as stationary either with a constant like Div5 or without a constant like Div2, Div3, Div6, Div7, Div8. The evidence concerning national accounting data is mixed. Recalling that the random walk with drift hypothesis cannot be rejected, it seems that at least Na1, Na2, Na3 can reasonably be decribed as following a random walk with trend while the remaining processes Na4, Na5, Na6 appears to be best represented by a more complex ARIMA–process than the analysis applied here is able to figure out. Moreover, Na6 may probably contain a constant in addition. Most of the proxy variables are likely to follow a pure random walk, except for Pi4, Io2 and It2, additionally exhibiting significant trend components. Moreover, some processes may better be represented by a more complex ARIMA–process like Pi1, Pi3, Io1, It1, It7. Notice that these processes are precisely those which have been identified as including a drift component in the preliminary regression[51]. Besides these non–stationary processes,

[50]Notice that the preliminary regression yielded no significant constant when testing for the random walk with drift hypothesis.

[51]This leads to the conclusion that the drift component in equation (5.11) captures

Pi2 appears to show stationary behaviour containing trend and constant, although the evidence is mixed[52]. Summarizing, most of the exogenous processes can be regarded as following a random walk, except for Pi2 and most dividend processes .

As already mentioned before, when testing directly for the geometric intrinsic bubble II (G4 in table 2.8 of section 2.4)[53], fundamental processes are required which do not reject the null hypothesis of a unit root in logs. Hence, by applying the same procedure described above to all fundamental processes in logs, we aim to gain evidence that the fundamental processes can be described as a *geometric random walk*. The results are reported in appendix A.5.4. The outcome of a preliminary regression is not reported here since it does not differ significantly from that obtained in the preceding subsection. Hence, the same caveats concerning critical values mentioned there are valid here.

App. A.5.4: The results concerning *fundamentals in logs* can be interpreted as follows: Concerning dividend processes it becomes obvious that the stationarity of dividends in levels partly vanishes when dividends in logs are considered. Here, besides Div1, the dividend processes Div6, Div7 can be regarded as pure random walks while the evidence for Div4 is mixed. Further investigation shows that Div4 should be regarded as stationary. Div2, Div3, Div5, Div8 remain stationary in comparison to the analysis conducted for processes in levels. Div2 includes trend and constant, Div5 only a constant while Div3, Div8 are best described as pure AR–processes. Within the set of national accounting data, Na1, Na3, Na5 appear non–stationary with trend while N2, Na4, Na6 follow a more complex ARIMA process[54]. Concerning the class of proxy variables it is easily be seen that no great differences are obtained in comparison to the analysis

the more complex structures not covered by simple random walk with drift models.

[52]Notice that the significance of the constant has already been shown by the preliminary regression.

[53]See Froot, Obstfeld (1991b).

[54]Comparing these results with the outcome for processes in levels it turns out that Na–data either contain a trend and show simultaneously non–stationary behaviour or can be described by a more sophisticated ARIMA–process.

Table 5.5: Summary of integration tests

	Endogenous Process	Exogenous Process in levels	Exogenous Process in logs
Random walk with trend	[In1], In2, In3, Se1, Se4, Se5, Se9 [St2], St3, St4, [St5], St6, St7	Na1, Na2, Na3, Pi4, Io2, It2	Na1, Na3, Na5 Pi4, Io2 It2
Random walk with constant	Se6		
Pure random walk	Se2, Se3, Se7, [Se8], B10,	Div1, Pi5, Pi6, Pi7, Pi8, Pi9, Io3, Io4, Io5, Io6, Io7, It3, It4, It5, It6	Div1, Div6, Div7, [Pi5], Pi6, Pi7, Pi8, Pi9, [Io3], [Io4], Io5, [Io6], Io7, It4, It5, [It6]
Stationary Process with trend		[Pi2]	Div2, Pi2
Stationary Process with constant	St1	Div5	Div5
Pure stationary Process		Div2, Div3, Div4, Div6, Div7, Div8	[Div3], [Div4], Div8
ARIMA-Process		Na4, Na5, [Na6], [Pi1], [Pi3], [Io1], [It1], It7	Na2, Na4, Na6 [Pi1], [Pi3], [Io1], [It1], [It3], It7

Remarks: Processes encapsulated by [] means 'mixed evidence'.

carried out above[55]. Summarizing, most of the processes can be regarded as following a geometric random walk and can, therefore, be used in testing for permanent intrinsic bubbles. The results discussed thus far are summarized in table 5.5. It turns out that most of the processes considered are non–stationary, eventually including a trend or drift component. In particular, when testing for the permanent presence of an intrinsic bubble, requiring fundamentals following a geometric random walk, only sets 4, 5, 6, 7, 10, 39 should be excluded, due to their apparent stationarity.

[55]The only difference is that the evidence concerning non–stationarity becomes lower for some processes.

Chapter 6

Concluding Remarks

In the preceding chapters we dealt with the notion of rational bubbles. In chapter 2 the theoretical basis for the occurrence of bubbles in rational expectations models was laid by providing solution procedures to a simple stochastic difference equation. Particular and complementary solution components could be distinguished. Concerning the particular solution, forward and backward solutions were derived. When considering different fundamental processes explicitly, ARMA solutions and random walk solutions were discussed. The resulting forward solutions are summarized in table 2.1 of section 2.3. While the forward solution is usually interpreted as the fundamental solution (or the present value relation), the complementary solution, containing a martingale process, represents the bubble solution component. Based on the notion of a generalized stochastic martingale (GSTM) process, serving as a common root for rational bubbles, the set of bubbles could be enlarged significantly. Mainly, three different bubble types were distinguished: Markovian, intrinsic and extrinsic bubbles. The resulting bubble processes are summarized in table 2.8 of section 2.4.

Chapter 3 discusses the economic content of that simple stochastic difference equation and simultaneously the relevance of rational bubbles in comparison to the fundamental solution. Three economic applications were presented: first, the occurrence of rational bubbles on capital markets where additionally a comprehensive description of a so called bubble cycle was provided, second, the applicability of the stochastic difference equation on foreign exchange markets and, finally, Cagan's hyperinflation model where rational bubbles explain the explosiveness of the price level. It came out that rational bubbles are capable of theoretically capturing two main aspects beyond the traditional fundamental evaluation of asset prices: the psychological effect in the sense of self-fulfilling prophecies, inducing a, fundamentally not justified, trend behaviour of the endogenous process (often followed by a sharp incline in prices), and the effect of overreaction to (fundamental) news.

Chapter 4 describes various testing strategies applicable for gaining evidence for the presence of rational bubbles. In general, direct and indirect test procedures are distinguished. Concerning the indirect test procedure, variance bounds tests, closely related to the present value relation (i.e. the efficient market hypothesis), specification tests and integration and cointegration tests were presented. However, it turns out that indirect tests often have low power in confirming the existence of rational bubbles since any other factor which induces prices to deviate from its fundamentals, could principally not be distinguished from rational bubbles. Direct tests, however, focus more directly on the presence of rational bubbles. Two direct tests were discussed: a deterministic bubble during German hyperinflation and intrinsic bubbles on the US-stock market. However, a problem similar to the problem discussed in the context of indirect testing procedures arises. One must be aware of the possibility of unobserved variables when testing for rational bubbles directly. If these neglected variables are taken into account by the market participants when evaluating an asset, but omitted by the econometrician, a bubble-like behaviour could falsly be attributed to rational bubbles, while a change in fundamentals has to be considered as an appropriate reason. In face of this difficulty one must check carefully the fundamental process when calculating the underlying fundamental solution. However, it must be emphasized that this problem is a problem of the model's misspecification and meets principally every economic model built for empirical investigation. One must bear this fact in mind and continuously improve the structure of the model and the testing procedures.

Finally, chapter 5 directly tests for the existence of three bubble types previously derived in chapter 2 on the German stock market by using indexes, sector indexes and single stocks as endogenous processes and dividends, national accounting data and proxy variables as fundamentals. The results of chapter 5 can be summarized as follows.

- First, a temporary Markovian bubble was considered. Examining the temporary existence during two periods (1980–1987 and 1987–1990), it turned out that indeed a Markovian bubble has explanatory power for describing the time path behaviour of stock indexes within these periods. For single stocks, however, the evidence is lower.

- Second, a temporary intrinsic bubble is investigated next. For most sets results as similar to the case of Markovian bubbles were obtained, due to the similar nature of the intrinsic bubble specification considered. However, the explanatory power for the second bubble specification if measured in terms of the coefficient of determination, decreases. This may support the view that a fundamentally not justified trend behaviour is more likely to exist on the German stock market than an overreaction effect.

- Finally, the presence of a permanent intrinsic bubble was tested for. The analysis did not show a significant impact of this bubble specification on stock prices. Hence, the hypothesis of a permanent existence of intrinsic bubbles can be rejected. This result again supports the opinion that an overreaction effect is unlikely to be present in the data examined.

- Besides these interpretations concerning the bubble terms, it can be stated that in general the coefficients of the fundamental solution component are highly significantly estimated. Especially, the gross capital income, containing dividend payments as a component, performs well as a proxy for the cash flow underlying the stock price indexes. Moreover, in general the regression results when using indexes rather than single stocks, are more affirmative. This is mainly due to the fact that the fundamental processes used in our analysis (aggregates like national accounting data or other proxy variables) rather represent fundamentals related to aggregated stock indexes.

These fundamentals may, however, be inappropriate when related to single stocks.

However, alternatives to the concept of rational bubbles must be discussed seriously. Generally, we have discussed these at the end of each chapter. Figure 6.1 summarizes the genesis of the rational bubble approach by providing the assumption structure underlying the model, as well as the alternatives. However, it must be stressed that it is very unlikely, as

Figure 6.1: Genesis of the rational bubble model

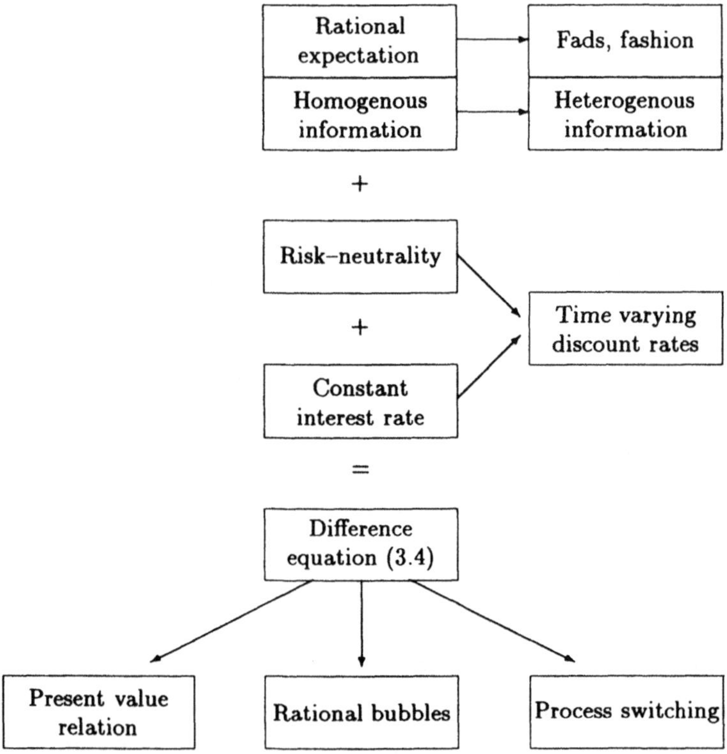

pointed out above, that these alternatives to equation (3.4) could describe the excessive deviation of the endogenous process from its fundamentals[1].

[1]Only the assumption of heterogenous information may have the potential to explain the observed price behaviour, however, lacks a theoretical framework to be tested for empirically.

Thus, it appears that they can only improve the evaluation of the fundamental solution while the explosive behaviour of the asset prices and crash phenomena are not explainable. On the other side, the search for eventually neglected fundamental factors or a possible process switching may be more promising. However, no such fundamentals have been found or even proven yet. As already mentioned, one must be aware of this possibility. However, at the present time, no such alternative model exists which is capable of explaining theoretically and proving empirically the phenomena observed. Thus, the existence of rational bubbles is the most obvious and convincing interpretation of the phenomena observed within these periods.

To conclude, the rational bubble model provides a theoretical framework for empirically examining psychological factors on asset markets. Any further research on rational bubbles should aim to enlarge the set of economically convincing bubble processes. Additionally, the search for more elaborate empirical approaches must be intensified if we wish to predict the excessive evolution of asset prices rather than only be able to confirm their existence ex–post.

Appendix A

Results

A.1 Temporary markovian bubbles

	$\hat{a}$	$\hat{b}$	$\hat{c}_0$	$\hat{c}_1$	$\hat{c}_2$	$\hat{c}_3$	$\hat{c}_4$
	DW	R^2	S1	E1	S2	E2	
Class 3							
Set 11	−334.294	3.014	1.171	1.731	179.773	1.946	155.239
	(−4.079)	(2.836)	(5.277)	(130.894)	(11.021)	(82.408)	(10.239)
	0.75	0.97	4:83	2:86	2:88	2:90	
Set 12	−140.747	4.239	3.458	1.809	155.882	2.289	108.515
	(−4.714)	(5.590)	(7.522)	(166.369)	(11.643)	(46.141)	(6.132)
	0.82	0.97	1:84	2:86	4:88	2:90	
Set 13	−345.433	3.012	1.290	1.730	177.069	1.938	152.346
	(−4.150)	(2.955)	(5.368)	(136.413)	(11.183)	(79.594)	(10.000)
	0.75	0.97	4:83	2:86	2:88	2:90	
Set 14	−284.753	1.474	1.304	1.616	230.416	2.196	181.936
	(−2.663)	(0.976)	(4.423)	(140.759)	(10.868)	(82.148)	(9.896)
	0.72	0.95	2:83	2:86	3:88	2:90	
Set 15	−100.784	2.265	4.301	1.600	206.708	1.812	75.364
	(−2.859)	(2.544)	(7.731)	(278.644)	(13.094)	(76.835)	(4.391)
	0.94	0.96	2:83	2:86	2:88	3:90	
Set 16	−301.104	1.421	1.448	1.615	227.741	2.186	178.814
	(−2.783)	(0.981)	(4.554)	(146.381)	(11.115)	(81.062)	(9.819)
	0.73	0.95	2:83	2:86	3:88	2:90	
Set 17	−114.352	0.753	0.550	1.727	100.835	1.964	82.550
	(−2.386)	(1.114)	(4.167)	(103.386)	(10.277)	(72.400)	(10.366)
	0.68	0.95	1:84	2:86	3:88	2:90	
Set 18	−44.960	0.940	1.930	1.709	91.393	1.657	37.519
	(−3.181)	(2.642)	(8.723)	(213.833)	(12.703)	(77.232)	(5.259)
	0.96	0.96	1:84	2:86	2:88	3:90	
Set 19	−125.679	0.665	0.623	1.729	100.472	1.800	81.870
	(−2.665)	(1.059)	(4.504)	(112.786)	(10.743)	(80.781)	(10.434)
	0.72	0.95	1:84	2:86	2:88	2:90	
Class 4							

Remarks: Values reported are the estimates of the coefficients in the regression equation (5.2). The t–statistics are presented below the estimates in parenthesis. The results refer to non–linear OLS–regression using quarterly data deflated by PPI. The variables 'S1' and 'S2' denote the starting point, the variables 'E1' and 'E2' the stopping time of the corresponding deterministic bubble. '4:83' then means the fourth quarter of 1983.

	$\hat{a}$	$\hat{b}$	$\hat{c}_0$	$\hat{c}_1$	$\hat{c}_2$	$\hat{c}_3$	$\hat{c}_4$
	DW	R^2	S1	E1	S2	E2	
Set 20	−183.873	1.097	1.643	1.461	69.511	1.595	46.074
	(−4.377)	(2.713)	(6.251)	(169.117)	(13.706)	(84.673)	(6.066)
	0.59	0.95	2:83	2:86	1:88	3:90	
Set 21	−217.431	0.971	1.857	1.351	140.925	1.633	65.787
	(−4.216)	(2.162)	(6.019)	(292.590)	(20.749)	(95.943)	(8.976)
	0.79	0.95	1:82	2:86	1:88	3:90	
Set 22	−83.277	−0.066	3.277	3.710	60.052	1.751	96.782
	(−2.403)	(−0.185)	(5.395)	(29.036)	(14.044)	(85.909)	(15.895)
	0.61	0.91	4:85	3:86	4:87	4:89	
Set 23	−103.936	1.483	1.190	1.653	68.381	2.099	43.354
	(−1.788)	(2.947)	(3.383)	(86.743)	(9.471)	(18.182)	(4.431)
	0.41	0.91	1:84	2:86	4:87	1:89	
Set 24	−101.685	−0.905	1.284	1.561	23.333	1.583	59.912
	(−4.073)	(−5.237)	(8.738)	(124.538)	(4.609)	(103.719)	(10.633)
	0.74	0.83	1:84	2:86	1:88	3:90	
Set 25	−133.470	2.986	1.373	1.495	260.701	1.593	65.724
	(−1.455)	(4.016)	(2.555)	(202.138)	(16.995)	(80.133)	(4.717)
	0.68	0.92	4:82	2:86	4:87	3:90	
Set 26	−167.632	−0.921	5.020	2.847	88.062	1.800	114.205
	(−3.397)	(−1.429)	(5.850)	(23.150)	(12.359)	(85.929)	(13.960)
	0.90	0.90	3:85	3:86	4:87	4:89	
Set 27	−9.601	−1.043	1.522	3.257	122.288	2.588	82.025
	(−0.334)	(−0.882)	(3.458)	(36.132)	(7.336)	(44.013)	(8.602)
	0.52	0.89	1:85	1:86	1:89	2:90	
Set 28	−32.018	7.878	0.652	1.604	245.183	36.819	159.043
	(−0.562)	(3.982)	(0.828)	(142.441)	(6.749)	(1.801)	(8.266)
	0.82	0.95	2:83	3:86	2:89	3:89	
Class 5							
Set 29	−151.725	1.841	2.703	3.703	71.476	1.664	40.833
	(−2.484)	(3.642)	(3.824)	(23.464)	(11.708)	(82.681)	(6.043)
	0.38	0.92	2:85	1:86	2:88	3:90	
Set 30	−126.281	1.857	2.464	3.646	69.686	1.662	40.044
	(−2.129)	(3.471)	(3.522)	(22.248)	(11.311)	(86.398)	(6.215)
	0.37	0.92	2:85	1:86	2:88	3:90	
Set 31	−123.183	1.649	2.540	6.705	72.819	1.742	27.574
	(−2.692)	(3.294)	(4.490)	(17.137)	(13.557)	(76.609)	(4.147)
	0.45	0.93	3:85	1:86	3:88	3:90	
Set 32	−96.568	1.765	2.224	3.596	70.392	1.662	41.888
	(−1.566)	(2.828)	(2.939)	(21.191)	(10.468)	(85.202)	(6.515)
	0.37	0.92	2:85	1:86	2:88	3:90	
Set 33	−46.301	1.992	1.753	2.137	129.924	1.817	58.389
	(−1.016)	(3.869)	(3.278)	(65.334)	(19.155)	(96.365)	(11.444)
	0.49	0.92	3:84	1:86	3:88	3:90	
Set 34	−79.648	1.674	2.216	2.481	137.691	1.939	40.373
	(−2.016)	(3.718)	(4.766)	(64.305)	(21.693)	(85.440)	(6.443)
	0.62	0.93	4:84	1:86	4:88	3:90	
Set 35	−42.339	1.827	1.763	2.512	134.985	1.705	60.803
	(−0.925)	(3.289)	(3.136)	(59.284)	(18.480)	(96.222)	(9.839)
	0.52	0.92	4:84	1:86	2:88	3:90	
Set 36	−208.841	1.173	3.108	3.414	52.942	1.666	74.522
	(−4.740)	(4.237)	(6.456)	(27.851)	(14.010)	(72.958)	(13.978)
	0.55	0.93	4:85	3:86	4:87	4:89	
Set 37	−203.193	0.753	3.176	2.544	55.240	1.929	66.795
	(−5.841)	(2.513)	(7.849)	(36.913)	(14.460)	(56.190)	(13.335)
	0.65	0.94	3:85	3:86	2:88	4:89	

Remarks: Values reported are the estimates of the coefficients in the regression equation (5.2). The t–statistics are presented below the estimates in parenthesis. The results refer to non–linear OLS–regression using quarterly data deflated by PPI. The variables 'S1' and 'S2' denote the starting point, the variables 'E1' and 'E2' the stopping time of the corresponding deterministic bubble. '4:83' then means the fourth quarter of 1983.

	$\hat{a}$	$\hat{b}$	$\hat{c}_0$	$\hat{c}_1$	$\hat{c}_2$	$\hat{c}_3$	$\hat{c}_4$
	DW	R^2	S1	E1	S2	E2	
Set 38	−185.245	0.868	2.982	3.384	54.652	1.667	73.858
	(−4.414)	(2.696)	(6.310)	(24.841)	(12.608)	(68.345)	(13.313)
	0.52	0.92	4:85	3:86	4:87	4:89	
Set 40	−341.573	0.343	5.111	7.180	79.633	1.477	3.172
	(−5.430)	(0.803)	(6.784)	(26.650)	(19.568)	(63.564)	(0.387)
	0.52	0.93	3:85	1:86	4:87	3:90	
Set 41	−182.296	0.721	3.489	6.433	69.161	1.452	−4.270
	(−2.276)	(1.008)	(3.456)	(13.960)	(10.557)	(48.463)	(−0.407)
	0.37	0.91	3:85	1:86	4:87	3:90	
Set 42	−203.080	0.647	3.716	6.466	70.038	1.453	−0.359
	(−2.381)	(0.889)	(3.494)	(14.301)	(10.839)	(49.886)	(−0.037)
	0.38	0.91	3:85	1:86	4:87	3:90	
Set 43	−25.077	0.197	1.469	3.254	17.650	1.720	47.151
	(−0.568)	(0.679)	(3.134)	(21.863)	(3.795)	(71.482)	(8.353)
	0.44	0.72	3:85	2:86	3:88	3:90	
Set 44	−14.601	0.175	1.446	−6.784	24.675	2.055	45.445
	(−0.568)	(0.803)	(5.210)	(−1.625)	(5.866)	(47.199)	(6.996)
	0.39	0.74	3:85	4:85	1:89	3:90	
Set 45	−2.361	0.141	1.313	3.440	18.652	1.744	51.284
	(−0.050)	(0.444)	(2.460)	(21.282)	(3.562)	(68.133)	(7.174)
	0.43	0.70	3:85	2:86	3:88	3:90	
Set 46	−5.692	3.795	1.194	2.096	246.674	1.880	59.988
	(−0.095)	(4.975)	(1.602)	(88.137)	(17.401)	(66.922)	(4.720)
	0.60	0.91	2:84	1:86	3:88	3:90	
Set 47	1.083	3.556	1.363	2.366	250.218	1.571	40.420
	(0.018)	(4.705)	(1.687)	(75.704)	(19.738)	(76.833)	(1.813)
	0.69	0.92	3:84	1:86	4:87	3:90	
Set 48	48.555	4.022	0.673	2.103	253.304	1.591	68.503
	(1.146)	(5.786)	(1.295)	(91.984)	(18.318)	(91.346)	(6.199)
	0.60	0.91	2:84	1:86	4:87	3:90	
Set 49	−285.838	1.186	4.015	2.280	76.511	2.907	81.909
	(−4.129)	(2.594)	(5.206)	(35.959)	(12.719)	(45.107)	(12.616)
	0.54	0.88	3:85	4:86	4:88	4:89	
Set 50	−274.822	0.654	4.063	2.001	79.761	7.918	72.567
	(−3.959)	(1.254)	(5.149)	(41.346)	(12.226)	(21.650)	(9.950)
	0.60	0.88	2:85	4:86	2:89	4:89	
Set 51	−247.840	0.821	3.773	2.273	78.799	2.908	81.444
	(−3.638)	(1.503)	(4.819)	(32.928)	(11.257)	(41.653)	(12.140)
	0.50	0.87	3:85	4:86	4:88	4:89	
Class 6							
Set 52	242.013	−0.613	−0.400	2.054	54.954	6.666	48.263
	(1.872)	(−0.816)	(−0.297)	(13.499)	(4.071)	(11.649)	(4.018)
	0.21	0.18	4:86	3:87	2:89	4:89	
Set 53	153.468	−0.840	0.564	−1.447	43.113	−1.278	31.006
	(1.769)	(−1.221)	(0.595)	(−10.016)	(2.942)	(−1.912)	(1.943)
	0.21	0.19	3:85	1:87	2:88	3:89	
Set 54	285.659	−0.432	−0.917	2.038	52.981	14.958	53.310
	(2.225)	(−0.531)	(−0.643)	(11.743)	(3.846)	(1.654)	(4.992)
	0.22	0.20	2:86	3:87	2:89	3:89	
Set 55	−264.926	1.155	3.654	6.749	66.225	1.900	35.492
	(−3.864)	(3.441)	(4.700)	(20.416)	(12.878)	(22.892)	(3.574)
	0.42	0.92	3:85	1:86	4:87	3:89	
Set 56	−118.218	1.682	2.085	6.173	57.797	1.897	30.678
	(−1.455)	(2.517)	(2.065)	(12.251)	(7.974)	(20.740)	(2.810)
	0.34	0.91	3:85	1:86	4:87	3:89	

Remarks: Values reported are the estimates of the coefficients in the regression equation (5.2). The t–statistics are presented below the estimates in parenthesis. The results refer to non-linear OLS-regression using quarterly data deflated by PPI. The variables 'S1' and 'S2' denote the starting point, the variables 'E1' and 'E2' the stopping time of the corresponding deterministic bubble. '4:83' then means the fourth quarter of 1983.

	$\hat{a}$	$\hat{b}$	$\hat{c}_0$	$\hat{c}_1$	$\hat{c}_2$	$\hat{c}_3$	$\hat{c}_4$
	DW	R^2	S1	E1	S2	E2	
Set 57	−138.751 (−1.623)	1.579 (2.341)	2.321 (2.205)	6.193 (12.507)	58.492 (8.170)	1.897 (20.728)	30.606 (2.868)
	0.35	0.91	3:85	1:86	4:87	3:89	
Set 58	−355.882 (−4.543)	1.618 (3.452)	4.548 (4.955)	27.438 (6.173)	73.930 (13.302)	1.359 (21.760)	−26.811 (−2.448)
	0.47	0.93	4:85	1:86	4:87	3:90	
Set 59	−178.897 (−1.817)	2.226 (2.627)	2.664 (2.152)	18.514 (2.830)	63.763 (8.163)	1.319 (13.650)	−31.134 (−2.322)
	0.36	0.91	4:85	1:86	4:87	3:90	
Set 60	−193.412 (−1.805)	2.179 (2.475)	2.821 (2.116)	3.883 (23.488)	65.480 (8.091)	1.321 (14.131)	−28.058 (−2.254)
	0.37	0.91	3:85	2:86	4:87	3:90	
Set 61	149.382 (1.097)	9.569 (5.127)	−1.731 (−0.956)	1.310 (75.453)	217.506 (14.505)	1.426 (19.212)	−19.639 (−0.955)
	0.41	0.88	1:83	3:87	4:87	3:90	
Set 62	85.117 (0.715)	9.190 (4.953)	−1.006 (−0.592)	1.305 (64.288)	209.145 (10.456)	1.417 (13.154)	−14.067 (−0.355)
	0.37	0.88	1:83	3:87	4:87	3:90	
Set 63	145.202 (1.386)	9.684 (5.338)	−1.715 (−1.156)	1.325 (60.758)	210.503 (12.018)	1.411 (16.557)	−20.198 (−1.063)
	0.42	0.88	2:83	3:87	4:87	3:90	
Set 64	−783.527 (−1.297)	9.356 (4.515)	7.867 (1.519)	0.942 (0.050)	−144.806 (−1.354)	1.920 (64.327)	586.409 (8.070)
	0.41	0.83	3:81	2:87	1:88	3:90	
Set 65	−694.681 (−2.131)	9.313 (6.271)	7.009 (2.558)	1.033 (0.498)	−89.794 (−1.123)	2.050 (74.424)	565.318 (10.287)
	0.38	0.86	2:81	3:87	2:88	3:90	
Set 66	−317.541 (−1.458)	5.045 (1.219)	4.896 (1.901)	−29.132 (−0.359)	−28.941 (−0.342)	1.842 (38.221)	667.387 (8.663)
	0.41	0.82	3:85	4:85	4:87	2:90	
Set 67	−802.652 (−6.093)	3.947 (4.506)	9.318 (6.345)	1.825 (14.097)	98.610 (7.430)	1.836 (3.048)	104.025 (9.889)
	0.48	0.92	2:85	4:86	3:88	3:89	
Set 68	−741.525 (−5.629)	2.825 (2.843)	9.063 (6.053)	3.701 (11.291)	104.261 (7.715)	−1.479 (−7.350)	85.676 (7.307)
	0.53	0.93	1:86	4:86	1:88	3:89	
Set 69	−724.369 (−5.530)	3.061 (2.882)	8.862 (5.875)	1.802 (11.522)	104.514 (6.924)	1.907 (3.213)	102.697 (8.993)
	0.45	0.92	2:85	4:86	3:88	3:89	
Set 70	139.936 (0.960)	3.723 (3.860)	−0.949 (−0.600)	2.169 (6.386)	−40.671 (−3.129)	3.196 (25.985)	72.215 (3.999)
	0.25	0.73	3:85	3:86	1:89	2:90	
Set 71	43.605 (0.587)	3.495 (4.124)	0.067 (0.078)	2.064 (4.469)	−40.512 (−2.471)	9.404 (10.155)	55.396 (2.894)
	0.23	0.72	3:85	3:86	4:89	2:90	
Set 72	214.744 (1.454)	4.067 (3.946)	−1.854 (−1.091)	2.111 (5.031)	−41.551 (−3.219)	3.288 (25.528)	81.748 (4.439)
	0.26	0.74	3:85	3:86	2:89	2:90	

Remarks: Values reported are the estimates of the coefficients in the regression equation (5.2). The t–statistics are presented below the estimates in parenthesis. The results refer to non–linear OLS–regression using quarterly data deflated by PPI. The variables 'S1' and 'S2' denote the starting point, the variables 'E1' and 'E2' the stopping time of the corresponding deterministic bubble. '4:83' then means the fourth quarter of 1983.

A.2 Temporary intrinsic bubbles

	$\hat{a}$	$\hat{b}$	$\hat{c}_0$	$\hat{c}_1$	$\hat{c}_2$	$\hat{c}_3$	$\hat{c}_4$
	DW	R^2	S1	E1	S2	E2	
Class 3							
Set 11	−395.095 (−5.433)	2.304 (2.416)	1.329 (6.735)	1.551 (133.931)	0.367 (10.140)	1.814 (61.945)	0.263 (9.924)
	0.77	0.94	1:84	2:86	4:88	2:90	
Set 12	−165.497 (−7.016)	3.647 (6.924)	3.860 (11.275)	1.707 (214.983)	1.434 (11.544)	1.828 (60.530)	0.461 (4.297)
	0.91	0.94	4:83	2:86	3:88	3:90	
Set 13	−419.071 (−5.805)	2.175 (2.438)	1.495 (7.162)	1.493 (159.060)	0.392 (10.251)	1.827 (59.460)	0.280 (9.795)
	0.78	0.94	4:83	2:86	4:88	2:90	
Set 14	−396.258 (−4.069)	0.246 (0.178)	1.589 (5.916)	1.422 (153.884)	0.477 (10.011)	2.121 (58.162)	0.310 (9.262)
	0.70	0.89	2:83	2:86	1:89	2:90	
Set 15	−131.309 (−4.484)	1.674 (2.446)	4.762 (10.515)	1.544 (347.544)	1.889 (12.678)	2.047 (59.899)	0.467 (4.022)
	1.01	0.90	1:83	2:86	4:88	3:90	
Set 16	−407.052 (−4.149)	0.283 (0.215)	1.740 (6.043)	1.430 (160.074)	0.504 (10.226)	2.136 (57.465)	0.327 (9.299)
	0.71	0.90	2:83	2:86	1:89	2:90	
Set 17	−160.444 (−3.754)	0.217 (0.367)	0.669 (5.716)	1.529 (108.184)	0.209 (9.800)	1.641 (61.762)	0.141 (9.883)
	0.70	0.90	2:84	2:86	4:88	2:90	
Set 18	−58.327 (−4.694)	0.707 (2.466)	2.125 (11.079)	1.620 (272.667)	0.834 (12.183)	1.825 (59.438)	0.234 (4.739)
	1.03	0.90	4:83	2:86	4:88	3:90	
Set 19	−175.908 (−4.179)	0.114 (0.206)	0.762 (6.191)	1.469 (128.215)	0.224 (9.955)	1.654 (60.123)	0.150 (9.859)
	0.71	0.90	1:84	2:86	4:88	2:90	
Class 4							
Set 20	−217.381 (−5.317)	0.892 (2.334)	1.838 (7.182)	1.349 (190.035)	0.343 (13.419)	1.614 (67.454)	0.190 (5.707)
	0.58	0.91	1:83	2:86	3:88	3:90	
Set 21	−250.043 (−4.752)	0.760 (1.762)	2.039 (6.427)	1.301 (295.137)	0.689 (19.333)	1.662 (76.571)	0.272 (8.199)
	0.74	0.92	1:82	2:86	3:88	3:90	
Set 22	−93.890 (−2.746)	−0.104 (−0.284)	3.435 (5.728)	2.963 (35.723)	0.850 (13.506)	1.808 (80.535)	1.155 (13.967)
	0.63	0.82	3:85	3:86	4:87	4:89	
Set 23	−160.571 (−2.896)	1.191 (2.941)	1.520 (4.636)	1.419 (121.216)	0.330 (9.869)	1.367 (44.574)	0.007 (0.130)
	0.41	0.85	3:83	2:86	4:87	3:90	
Set 24	−103.262 (−4.741)	−0.924 (−6.172)	1.296 (10.131)	1.520 (121.165)	0.113 (4.500)	1.602 (82.739)	0.255 (10.737)
	0.78	0.73	2:84	2:86	3:88	3:90	
Set 25	−200.488 (−2.125)	2.628 (3.554)	1.747 (3.143)	1.393 (222.144)	1.284 (16.144)	1.476 (78.373)	0.271 (4.278)
	0.65	0.91	3:82	2:86	4:87	3:90	

Remarks: Values reported are the estimates of the coefficients in the regression equation (5.5). The t–statistics are presented below the estimates in parenthesis. The results refer to non–linear OLS–regression using quarterly data deflated by PPI. The variables 'S1' and 'S2' denote the starting point, the variables 'E1' and 'E2' the stopping time of the corresponding intrinsic bubble. '4:83' then means the fourth quarter of 1983.

	$\hat{a}$	$\hat{b}$	$\hat{c}_0$	$\hat{c}_1$	$\hat{c}_2$	$\hat{c}_3$	$\hat{c}_4$
	DW	R^2	S1	E1	S2	E2	
Set 26	−199.406 (−4.643)	−1.164 (−1.844)	5.538 (6.732)	4.694 (23.082)	1.210 (11.055)	1.905 (49.832)	1.386 (12.609)
	0.80	0.79	3:85	2:86	4:87	3:89	
Set 27	−77.736 (−2.171)	−3.139 (−2.632)	2.446 (4.942)	1.736 (89.211)	0.975 (7.136)	2.080 (47.396)	0.540 (8.502)
	0.62	0.76	1:85	2:86	4:88	2:90	
Set 28	−147.163 (−2.207)	4.384 (2.245)	2.193 (2.572)	1.449 (175.765)	1.962 (6.645)	43.029 (3.015)	1.090 (8.564)
	0.87	0.91	3:82	3:86	2:89	3:89	
Class 5							
Set 29	−181.436 (−2.865)	1.676 (3.236)	3.041 (4.124)	3.705 (24.686)	0.706 (12.355)	1.634 (75.967)	0.329 (5.038)
	0.33	0.87	2:85	1:86	2:88	3:90	
Set 30	−157.564 (−2.580)	1.664 (3.048)	2.827 (3.900)	3.642 (23.092)	0.687 (11.809)	1.630 (77.011)	0.316 (4.938)
	0.33	0.87	2:85	1:86	2:88	3:90	
Set 31	−154.511 (−3.365)	1.410 (2.768)	2.926 (5.096)	6.647 (17.217)	0.739 (13.397)	1.828 (56.276)	0.188 (2.960)
	0.39	0.87	3:85	1:86	4:88	3:90	
Set 32	−137.138 (−2.116)	1.473 (2.315)	2.716 (3.413)	6.808 (15.992)	0.702 (11.006)	1.728 (64.820)	0.316 (4.747)
	0.33	0.84	3:85	1:86	3:88	3:90	
Set 33	−61.311 (−1.191)	1.841 (3.335)	1.956 (3.205)	2.123 (62.182)	1.233 (18.799)	2.141 (63.172)	0.453 (7.710)
	0.39	0.84	3:84	1:86	1:89	3:90	
Set 34	−97.336 (−2.179)	1.497 (3.048)	2.452 (4.603)	2.466 (62.159)	1.389 (22.075)	2.079 (58.976)	0.286 (4.469)
	0.50	0.87	4:84	1:86	1:89	3:90	
Set 35	−65.542 (−1.197)	1.601 (2.565)	2.067 (3.040)	2.501 (58.213)	1.322 (18.418)	1.775 (77.958)	0.466 (7.476)
	0.43	0.85	4:84	1:86	3:88	3:90	
Set 36	−210.593 (−4.517)	1.169 (4.172)	3.125 (6.145)	3.387 (28.505)	0.513 (14.399)	1.644 (71.851)	0.644 (13.281)
	0.54	0.90	4:85	3:86	4:87	4:89	
Set 37	−210.202 (−5.436)	0.721 (2.343)	3.251 (7.287)	2.537 (38.451)	0.538 (15.046)	1.881 (52.955)	0.553 (11.823)
	0.63	0.89	3:85	3:86	2:88	4:89	
Set 38	−190.828 (−4.236)	0.845 (2.552)	3.041 (5.989)	3.362 (25.692)	0.534 (13.027)	1.642 (67.126)	0.624 (12.522)
	0.50	0.87	4:85	3:86	4:87	4:89	
Set 40	−353.130 (−6.061)	0.228 (0.520)	5.261 (7.373)	7.214 (26.988)	0.800 (20.201)	1.464 (64.479)	0.032 (0.458)
	0.52	0.89	3:85	1:86	4:87	3:90	
Set 41	−222.835 (−3.214)	0.282 (0.429)	4.035 (4.578)	6.444 (14.007)	0.703 (10.903)	1.453 (47.169)	−0.044 (−0.501)
	0.35	0.81	3:85	1:86	4:87	3:90	
Set 42	−244.997 (−3.402)	0.222 (0.340)	4.271 (4.725)	6.482 (14.391)	0.712 (11.404)	1.436 (48.493)	−0.006 (−0.075)
	0.36	0.82	3:85	1:86	4:87	3:90	
Set 43	−20.540 (−0.464)	0.198 (0.673)	1.426 (3.043)	3.201 (21.577)	0.167 (3.802)	1.812 (59.481)	0.388 (7.968)
	0.42	0.43	3:85	2:86	4:88	3:90	
Set 44	−12.942 (−0.507)	0.139 (0.609)	1.441 (5.125)	−6.741 (−1.589)	0.258 (6.071)	1.998 (45.256)	0.386 (6.605)
	0.37	0.44	3:85	4:85	1:89	3:90	

Remarks: Values reported are the estimates of the coefficients in the regression equation (5.5). The t–statistics are presented below the estimates in parenthesis. The results refer to non–linear OLS–regression using quarterly data deflated by PPI. The variables 'S1' and 'S2' denote the starting point, the variables 'E1' and 'E2' the stopping time of the corresponding intrinsic bubble. '4:83' then means the fourth quarter of 1983.

	$\hat{a}$	$\hat{b}$	$\hat{c}_0$	$\hat{c}_1$	$\hat{c}_2$	$\hat{c}_3$	$\hat{c}_4$
	DW	R^2	S1	E1	S2	E2	
Set 45	5.992	0.132	1.232	3.477	0.188	1.715	0.445
	(0.126)	(0.408)	(2.296)	(21.425)	(3.618)	(66.262)	(7.068)
	0.40	0.41	3:85	2:86	3:88	3:90	
Set 46	−4.005	3.689	1.334	2.090	2.301	2.242	0.469
	(−0.062)	(4.920)	(1.671)	(83.818)	(15.389)	(44.824)	(3.716)
	0.50	0.85	2:84	1:86	1:89	3:90	
Set 47	−76.688	2.964	2.409	2.332	2.429	3.064	0.056
	(−1.074)	(4.097)	(2.549)	(58.219)	(18.796)	(16.568)	(0.239)
	0.60	0.86	3:84	1:86	3:89	3:90	
Set 48	30.642	3.865	0.902	2.099	2.448	1.563	0.526
	(0.618)	(5.474)	(1.470)	(90.072)	(17.378)	(83.335)	(4.763)
	0.50	0.87	2:84	1:86	4:87	3:90	
Set 49	−314.182	1.119	4.292	1.990	0.760	1.608	0.691
	(−4.332)	(2.361)	(5.263)	(44.232)	(13.254)	(26.611)	(11.550)
	0.53	0.83	2:85	4:86	4:87	4:89	
Set 50	−307.674	0.537	4.405	2.004	0.787	7.294	0.585
	(−4.194)	(0.997)	(5.252)	(44.033)	(13.065)	(18.589)	(8.219)
	0.59	0.81	2:85	4:86	2:89	4:89	
Set 51	−281.964	0.685	4.128	2.283	0.791	1.612	0.671
	(−3.902)	(1.193)	(4.917)	(35.268)	(11.881)	(24.514)	(10.606)
	0.48	0.79	3:85	4:86	4:87	3:89	
Class 6							
Set 52	254.396	−0.576	−0.462	−47.753	−0.195	−1.060	0.337
	(1.849)	(−0.861)	(−0.335)	(−2.026)	(−0.927)	(−1.350)	(2.817)
	0.29	0.30	1:84	2:84	1:88	3:89	
Set 53	111.959	−0.985	1.042	−1.489	0.498	−15.192	0.186
	(1.149)	(−1.404)	(0.974)	(−10.636)	(3.192)	(−0.793)	(1.083)
	0.21	0.33	3:85	1:87	2:89	3:89	
Set 54	284.311	−0.468	−0.823	−54.338	−0.184	6.015	0.398
	(2.025)	(−0.628)	(−0.547)	(−2.036)	(−0.907)	(8.244)	(3.225)
	0.30	0.37	2:84	3:84	2:89	4:89	
Set 55	−265.911	1.070	3.685	6.814	0.669	2.346	0.342
	(−4.662)	(3.486)	(5.648)	(20.184)	(12.919)	(12.848)	(4.030)
	0.44	0.90	3:85	1:86	4:87	4:88	
Set 56	−148.148	1.314	2.496	6.234	0.593	2.326	0.297
	(−2.180)	(2.290)	(2.961)	(12.151)	(8.117)	(11.095)	(3.133)
	0.33	0.85	3:85	1:86	4:87	4:88	
Set 57	−169.205	1.224	2.733	6.258	0.600	2.344	0.298
	(−2.436)	(2.167)	(3.205)	(12.450)	(8.381)	(11.587)	(3.206)
	0.34	0.86	3:85	1:86	4:87	4:88	
Set 58	−368.275	1.480	4.713	27.884	0.744	1.352	−0.230
	(−5.446)	(3.245)	(5.836)	(6.185)	(13.680)	(23.023)	(−2.452)
	0.46	0.91	4:85	1:86	4:87	3:90	
Set 59	−226.860	1.714	3.308	18.648	0.651	1.305	−0.277
	(−2.818)	(2.340)	(3.250)	(2.842)	(8.546)	(13.507)	(−2.478)
	0.33	0.87	4:85	1:86	4:87	3:90	
Set 60	−244.122	1.672	3.490	3.905	0.670	1.306	−0.248
	(−2.839)	(2.253)	(3.253)	(23.854)	(8.656)	(13.934)	(−2.329)
	0.34	0.87	3:85	2:86	4:87	3:90	
Set 61	65.075	8.749	−0.605	1.320	1.956	2.253	−0.263
	(0.627)	(5.575)	(−0.437)	(64.577)	(13.045)	(4.508)	(−1.427)
	0.34	0.84	2:83	3:87	3:89	3:90	
Set 62	−13.811	8.084	0.409	1.323	2.114	1.373	−0.365
	(−0.122)	(5.088)	(0.257)	(56.745)	(10.560)	(3.820)	(−1.034)
	0.31	0.83	2:83	3:87	2:88	3:90	

Remarks: Values reported are the estimates of the coefficients in the regression equation (5.5). The t–statistics are presented below the estimates in parenthesis. The results refer to non-linear OLS–regression using quarterly data deflated by PPI. The variables 'S1' and 'S2' denote the starting point, the variables 'E1' and 'E2' the stopping time of the corresponding intrinsic bubble. '4:83' then means the fourth quarter of 1983.

	$\hat{a}$	$\hat{b}$	$\hat{c}_0$	$\hat{c}_1$	$\hat{c}_2$	$\hat{c}_3$	$\hat{c}_4$
	DW	R^2	S1	E1	S2	E2	
Set 63	58.628 (0.657)	8.829 (5.598)	−0.563 (−0.449)	1.320 (59.889)	1.983 (11.811)	2.201 (3.940)	−0.273 (−1.461)
	0.33	0.84	2:83	3:87	3:89	3:90	
Set 64	−832.801 (−1.619)	9.112 (4.205)	8.365 (1.841)	0.984 (0.126)	−1.290 (−1.547)	2.022 (61.505)	4.773 (8.381)
	0.41	0.80	4:81	2:87	2:88	3:90	
Set 65	−660.907 (−1.993)	8.872 (5.473)	6.845 (2.429)	1.009 (0.306)	−0.883 (−1.068)	1.992 (72.294)	4.470 (10.022)
	0.37	0.81	2:81	2:87	2:88	3:90	
Set 66	−152.383 (−0.592)	5.621 (1.239)	3.481 (1.176)	−80.186 (−1.977)	−0.801 (−0.961)	1.784 (32.048)	4.836 (7.562)
	0.39	0.66	1:84	2:84	4:87	2:90	
Set 67	−828.623 (−6.047)	3.884 (4.338)	9.585 (6.218)	6.853 (8.677)	0.959 (7.883)	1.730 (2.597)	0.893 (8.836)
	0.46	0.90	2:86	4:86	3:88	3:89	
Set 68	−793.323 (−5.836)	2.582 (2.530)	9.621 (6.178)	3.742 (12.486)	1.035 (8.169)	−1.433 (−6.663)	0.697 (6.142)
	0.52	0.88	1:86	4:86	1:88	3:89	
Set 69	−775.013 (−5.595)	2.844 (2.578)	9.403 (5.849)	1.825 (13.407)	1.043 (7.319)	1.854 (3.023)	0.863 (7.592)
	0.42	0.87	2:85	4:86	3:88	3:89	
Set 70	186.910 (1.239)	3.735 (4.114)	−1.356 (−0.849)	−56.237 (−2.231)	−0.328 (−1.740)	3.052 (22.537)	0.608 (3.752)
	0.26	0.69	1:84	2:84	2:89	2:90	
Set 71	9.572 (0.126)	3.240 (4.188)	0.546 (0.642)	−40.069 (−2.557)	−0.304 (−1.820)	7.871 (6.278)	0.382 (1.861)
	0.23	0.66	1:84	2:84	4:89	2:90	
Set 72	237.109 (1.556)	3.996 (4.057)	−2.006 (−1.167)	−66.979 (−2.423)	−0.312 (−1.870)	3.145 (22.175)	0.690 (4.019)
	0.28	0.69	1:84	2:84	2:89	2:90	

Remarks: Values reported are the estimates of the coefficients in the regression equation (5.5). The t–statistics are presented below the estimates in parenthesis. The results refer to non–linear OLS–regression using quarterly data deflated by PPI. The variables 'S1' and 'S2' denote the starting point, the variables 'E1' and 'E2' the stopping time of the corresponding intrinsic bubble. '4:83' then means the fourth quarter of 1983.

A.3 Permanent intrinsic bubbles – Class 1 to 2

		Empirical Values						Theoretical Values				
		$\hat{c}_0$	$\hat{c}_1$	$\hat{\lambda}$	LR	DW	R^2	c_0	λ	r	μ	σ^2
Class 1												
Set 1	UR	9.137 (5.748)	$0.0^{[*]}$ (0.014)	11.978 (0.584)	3.72	0.28	0.21	5.725	6.208	0.167	$0.258^{[-2]}$	$0.781^{[-2]}$
	R	8.031 (3.600)	$0.200^{[-6]}$ (1.552)			0.28	0.21					
Set 2	UR	14.547 (10.262)	$0.0^{[*]}$ (0.040)	8.341 (1.096)	9.29	0.95	0.45	12.018	4.372	0.086	$0.258^{[-2]}$	$0.781^{[-2]}$
	R	13.025 (6.286)	$0.954^{[-4]}$ (1.884)			0.94	0.45					
Set 3	UR	6.315 (8.277)	$0.0^{[*]}$ (0.053)	6.469 (1.136)	12.79	1.02	0.52	6.049	6.051	0.159	$0.258^{[-2]}$	$0.781^{[-2]}$
	R	6.230 (9.396)	$0.200^{[-6]}$ (2.293)			1.02	0.52					
Class 2												
Set 4	UR	−3.043 (−1.340)	137.49 (15.732)	0.186 (2.584)	121.63	1.71	0.20	15.273	1.366	0.073	-0.107	0.236
	R	59.317 (9.141)	−16.113 (−7.889)			0.75	–					
Set 5	UR	−4.480 (−0.488)	74.975 (1.280)	0.315 (0.534)	48.73	0.36	0.08	4.050	4.476	0.199	-0.046	0.041
	R	9.291 (8.740)	$-0.349^{[-4]}$ (−5.402)			0.39	–					
Set 6	UR	−14.820 (−0.359)	62.007 (4.170)	0.624 (0.880)	35.54	0.23	0.10	4.104	3.894	0.207	-0.037	0.046
	R	10.556 (8.375)	$-0.369^{[-3]}$ (−5.420)			0.30	–					
Set 7	UR	−24.380 (−0.231)	160.051 (0.540)	0.416 (0.221)	18.77	0.32	0.14	3.776	1.837	0.331	0.048	0.143
	R	32.525 (5.527)	−2.268 (−4.142)			0.36	–					
Set 8	UR	−1.084 (−0.186)	$2.701^{[3]}$ (0.327)	−0.989 (−0.671)	36.91	0.29	0.29	2.805	6.704	0.290	-0.031	0.022
	R	17.351 (4.271)	$-0.313^{[-9]}$ (−3.165)			0.43	–					
Set 9	UR	−169.64 (−0.147)	249.81 (0.227)	0.899 (1.429)	65.62	0.69	0.20	7.743	4.277	0.105	-0.032	0.026
	R	25.226 (11.339)	$-0.234^{[-3]}$ (−7.318)			0.65	–					
Set 10	UR	−2.257 (−0.519)	211.49 (1.150)	−0.261 (−0.442)	55.80	0.73	0.32	3.068	3.921	0.266	-0.052	0.061
	R	9.360 (4.609)	$-0.691^{[-3]}$ (−2.999)			0.28	–					

Remarks: Values reported are the estimates of the coefficients in the regression equation (5.7). The t–statistics are presented below the estimates in parenthesis. The results refer to non–linear OLS–regression using quarterly data deflated by PPI. 'UR' corresponds to unrestricted regression and 'R' to restricted regression. Numbers encapsulated in brackets [] indicate the exponential degree to 10. [*] means estimate can not be distinguished from zero.

A.4 Permanent intrinsic bubbles – Class 3 to 6

	$\hat{c}_0$	$\hat{c}_1$	$\hat{\lambda}$	LR	DW	R^2
Class 3						
Set 11	1.131 (18.675)	−3.166[8] (−1.272)	−2.321 (−16.689)	71.91	0.23	0.83
Set 12	4.600 (7.522)	−5.117[6] (−0.085)	−2.361 (−0.855)	58.20	0.35	0.85
Set 13	1.255 (20.908)	−8.217[7] (−0.441)	−2.118 (−5.635)	73.68	0.23	0.84
Set 14	1.145 (14.223)	−6.293[7] (−0.180)	−2.120 (−2.285)	27.38	0.38	0.67
Set 15	4.699 (21.646)	−2.344[8] (−0.711)	−3.362 (−10.674)	21.27	0.50	0.74
Set 16	1.244 (13.811)	−9.135[7] (−0.330)	−2.205 (−4.291)	29.83	0.39	0.68
Set 17	0.522 (11.688)	−4.392[7] (−0.172)	−2.196 (−2.244)	31.13	0.20	0.73
Set 18	2.157 (18.442)	−4.028[7] (−0.122)	−3.165 (−1.642)	24.59	0.33	0.80
Set 19	0.566 (18.806)	−7.353[7] (−0.268)	−2.307 (−3.722)	33.94	0.20	0.74
Class 4						
Set 20	1.459 (7.702)	−3.438[7] (−0.151)	−2.423 (−1.826)	62.49	0.29	0.77
Set 21	1.504 (9.583)	−6.595[7] (−0.166)	−2.566 (−2.148)	27.94	0.21	0.57
Set 22	2.670 (12.377)	−1.827[8] (−0.631)	−3.696 (−8.365)	23.21	0.21	0.68
Set 23	1.393 (10.555)	−1.031[8] (−0.362)	−2.658 (−4.735)	50.35	0.23	0.73
Set 24	0.709 (5.554)	−3.189[6] (−0.013)	−2.520 (−0.160)	1.15	0.32	0.41
Set 25	2.081 (2.986)	−3.370[7] (−0.057)	−2.355 (−0.674)	18.07	0.17	0.44
Set 26	3.054 (13.449)	−7.159[8] (−1.636)	−3.965 (−21.099)	26.84	0.24	0.67
Set 27	1.451 (8.600)	−0.0[*] (−0.035)	−5.629 (−1.091)	12.36	0.20	0.59
Set 28	3.057 (13.126)	−1.015[9] (−0.916)	−3.444 (−14.989)	27.90	0.16	0.70
Class 5						
Set 29	3.250 (4.632)	−3.665[7] (−0.114)	−2.706 (−1.337)	67.86	0.27	0.73
Set 30	3.122 (7.701)	−4.388[7] (−0.263)	−2.769 (−3.114)	71.30	0.29	0.74
Set 31	2.888 (7.165)	−3.219[7] (−0.150)	−2.777 (−1.785)	68.13	0.31	0.76
Set 32	2.893 (9.157)	−4.727[7] (−0.236)	−2.854 (−2.882)	68.24	0.31	0.76

Remarks: Values reported are the estimates of the coefficients in the regression equation (5.7). The t–statistics are presented below the estimates in parenthesis. The results refer to non-linear OLS–regression using quarterly data deflated by PPI. Numbers encapsulated in brackets '[]' indicate the exponential degree to 10. '[*] means estimate can not be distinguished from zero.

	$\hat{c}_0$	$\hat{c}_1$	$\hat{\lambda}$	LR	DW	R^2
Set 33	2.950 (10.059)	$-6.480^{[7]}$ (−0.282)	−2.962 (−3.553)	41.19	0.28	0.65
Set 34	2.839 (13.601)	$-7.245^{[7]}$ (−0.235)	−3.041 (−3.068)	33.09	0.26	0.62
Set 35	2.818 (10.279)	$-6.752^{[7]}$ (−0.158)	−3.011 (−2.067)	34.45	0.29	0.62
Set 36	2.803 (1.940)	$-7.340^{[5]}$ (−0.051)	−2.354 (−0.522)	46.58	0.19	0.61
Set 37	2.571 (3.221)	$-9.349^{[6]}$ (−0.075)	−2.453 (−0.807)	71.87	0.32	0.75
Set 38	2.572 (2.516)	$-1.513^{[7]}$ (−0.054)	−2.563 (−0.607)	58.02	0.25	0.70
Set 40	3.102 (3.988)	$-1.617^{[7]}$ (−0.092)	−2.578 (−1.017)	76.78	0.29	0.78
Set 41	2.921 (2.091)	$-8.683^{[5]}$ (−0.050)	−1.998 (−0.423)	67.58	0.30	0.79
Set 42	2.919 (2.310)	$-2.185^{[6]}$ (−0.050)	−2.199 (−0.474)	69.81	0.30	0.79
Set 43	1.196 (4.645)	$-0.0^{[*]}$ (−0.017)	7.738 (0.645)	24.60	0.45	0.57
Set 44	1.320 (9.802)	$-0.0^{[*]}$ (−0.022)	7.822 (0.832)	15.37	0.45	0.56
Set 45	1.276 (5.137)	$-0.0^{[*]}$ (−0.013)	7.705 (0.471)	15.25	0.45	0.49
Set 46	3.888 (3.505)	$-2.360^{[7]}$ (−0.055)	−2.695 (−0.635)	32.48	0.39	0.54
Set 47	3.708 (9.917)	$-5.152^{[8]}$ (−0.758)	−3.494 (−11.055)	28.05	0.32	0.56
Set 48	3.611 (6.949)	$-5.588^{[7]}$ (−0.117)	−2.956 (−1.494)	22.22	0.32	0.48
Set 49	3.175 (2.243)	$-2.025^{[7]}$ (−0.058)	−2.547 (−0.652)	48.01	0.22	0.61
Set 50	2.873 (5.640)	$-3.489^{[7]}$ (−0.141)	−2.718 (−1.671)	66.22	0.33	0.71
Set 51	2.913 (4.086)	$-2.768^{[7]}$ (−0.102)	−2.668 (−1.186)	53.58	0.26	0.66
Class 6						
Set 52	4.108 (0.139)	−0.034 (−0.016)	1.911 (0.173)	17.91	0.24	0.01
Set 53	−14.074 (−0.014)	27.124 (0.034)	0.884 (0.120)	16.22	0.24	0.00
Set 54	−0.188 (−0.017)	$9.373^{[2]}$ (0.033)	−0.332 (−0.043)	25.74	0.25	0.04
Set 55	2.876 (2.134)	$-8.391^{[6]}$ (−0.068)	−2.396 (−0.698)	88.55	0.27	0.79
Set 56	2.959 (1.310)	$-6.004^{[4]}$ (−0.065)	−1.320 (−0.359)	83.40	0.27	0.80
Set 57	2.887 (1.510)	$-1.938^{[5]}$ (−0.065)	−1.587 (−0.430)	85.74	0.27	0.81
Set 58	3.846 (1.401)	$-5.062^{[5]}$ (−0.070)	−1.678 (−0.496)	102.52	0.26	0.82
Set 59	4.137 (0.905)	$-7.154^{[3]}$ (−0.079)	−0.734 (−0.232)	94.76	0.26	0.82

Remarks: Values reported are the estimates of the coefficients in the regression equation (5.7). The t-statistics are presented below the estimates in parenthesis. The results refer to non-linear OLS-regression using quarterly data deflated by PPI. Numbers encapsulated in brackets '[]' indicate the exponential degree to 10. '[*] means estimate can not be distinguished from zero.

	$\hat{c}_0$	$\hat{c}_1$	$\hat{\lambda}$	LR	DW	R^2
Set 60	3.981 (1.044)	$-1.786^{[4]}$ (−0.077)	−0.947 (−0.297)	97.00	0.26	0.82
Set 61	5.182 (2.046)	$-2.887^{[6]}$ (−0.063)	−2.068 (−0.556)	46.61	0.34	0.59
Set 62	4.503 (7.256)	$-3.830^{[8]}$ (−0.504)	−3.293 (−6.523)	39.97	0.25	0.59
Set 63	4.650 (3.425)	$-2.268^{[7]}$ (−0.075)	−2.586 (−0.839)	43.78	0.34	0.57
Set 64	−79.037 (−0.021)	38.838 (0.012)	1.160 (0.160)	29.61	0.18	0.42
Set 65	1.996 (0.816)	$0.0^{[*]}$ (0.021)	6.761 (0.728)	31.17	0.16	0.49
Set 66	7.140 (1.397)	$-2.691^{[3]}$ (−0.095)	−0.406 (−0.159)	31.12	0.20	0.62
Set 67	7.635 (1.316)	$-6.600^{[6]}$ (−0.054)	−2.046 (−0.482)	73.09	0.19	0.71
Set 68	6.578 (2.641)	$-1.902^{[7]}$ (−0.087)	−2.330 (−0.885)	113.24	0.39	0.83
Set 69	6.694 (2.699)	$-2.312^{[7]}$ (−0.089)	−2.370 (−0.927)	98.04	0.29	0.80
Set 70	2.798 (0.869)	$-1.382^{[6]}$ (−0.025)	−2.017 (−0.220)	12.64	0.11	0.26
Set 71	2.616 (1.757)	$-6.890^{[6]}$ (−0.037)	−2.466 (−0.386)	12.58	0.11	0.31
Set 72	2.727 (1.460)	$-1.022^{[7]}$ (−0.031)	−2.509 (−0.340)	17.41	0.15	0.31

Remarks: Values reported are the estimates of the coefficients in the regression equation (5.7). The t–statistics are presented below the estimates in parenthesis. The results refer to non–linear OLS–regression using quarterly data deflated by PPI. Numbers encapsulated in brackets '[]' indicate the exponential degree to 10. '[*] means estimate can not be distinguished from zero.

A.5 Integration tests

A.5.1 Preliminary regression of endogenous processes

	$\hat{m}$ ($t_{\hat{m}}$)	μ Lags		$\hat{m}$ ($t_{\hat{m}}$)	μ Lags		$\hat{m}$ ($t_{\hat{m}}$)	μ Lags
Indexes								
In1	5.436 (1.193)	0.004 2	In2	4.453 (0.823)	0.002 2	In3	1.834 (0.799)	0.005 2
Sector indexes								
Se1	1.958 (0.919)	0.007 3	Se2	1.675 (0.550)	0.003 3	Se3	0.545 (0.343)	0.004 5
Se4	1.421 (1.204)	0.017 2	Se5	2.579 (1.338)	0.011 3	Se6	−0.106 (−0.072)	−0.001 4
Se7	3.747 (0.732)	0.002 0	Se8	0.842 (0.380)	0.003 2	Se9	1.141 (0.483)	0.003 2
Se10	5.999 (0.988)	0.003 2						
Stocks								
St1	−0.815 (−0.335)	−0.002 0	St2	3.117* (1.775)	0.017 3	St3	3.564* (1.833)	0.016 3
St4	9.262* (1.666)	0.004 0	St5	6.400 (0.669)	0.001 2	St6	4.186 (1.016)	0.004 0
St7	2.780 (1.132)	0.008 3						

Remarks: Values reported are the coefficients in the regression of equation (5.11). The t-statistics, displayed in parentheses beneath the estimates, relate to the two-sided test for $m = 0$. Preliminary regression include the number of lagged differences of the variable noted in the table in order to correct for autocorrelation. Additionally, the theoretically derived value μ is given. The results refer to OLS-regression using quarterly data deflated by PPI.

* indicates significant values at 5%-level.

A.5.2 Preliminary regression of exogenous processes

	$\hat{m}$ $(t_{\hat{m}})$	μ Lags		$\hat{m}$ $(t_{\hat{m}})$	μ Lags		$\hat{m}$ $(t_{\hat{m}})$	μ Lags
Dividends								
Div1	0.001 (0.004)	0.000 3	Div2	−0.858* (−2.026)	−0.141 0	Div3	−0.698 (−1.207)	−0.061 0
Div4	−0.630 (−1.319)	−0.081 0	Div5	0.220 (0.335)	0.015 0	Div6	−0.572 (−1.036)	−0.055 0
Div7	−0.584 (−1.192)	−0.072 0	Div8	−0.547 (−1.061)	−0.061 0			
National accounting data								
Na1	3.392* (3.828)	0.103 2	Na2	1.006* (2.209)	0.072 0	Na3	3.962* (5.943)	0.133 0
Na4	0.917* (2.256)	0.083 0	Na5	0.506* (3.064)	0.392 0	Na6	0.622* (2.556)	0.501 2
Proxies								
Pi1	0.254 (1.468)	0.136 2	Pi2	0.482* (1.981)	0.122 0	Pi3	0.256 (1.389)	0.120 2
Pi4	0.262 (1.548)	0.159 3	Pi5	0.341 (1.344)	0.083 2	Pi6	0.180 (0.633)	0.034 2
Pi7	0.544 (0.896)	0.022 0	Pi8	0.090 (0.601)	0.061 2	Pi9	0.315 (0.563)	0.015 0
Io1	0.267 (1.225)	0.090 2	Io2	0.343* (1.806)	0.168 2	Io3	0.319 (1.157)	0.066 2
Io4	0.233 (0.646)	0.027 2	Io5	0.493 (0.896)	0.024 0	Io6	0.214 (0.941)	0.064 2
Io7	0.306 (0.654)	0.021 0						
It1	0.363* (1.648)	0.120 2	It2	0.345* (1.921)	0.192 2	It3	0.475 (1.608)	0.081 0
It4	0.185 (0.510)	0.021 0	It5	0.510 (0.767)	0.017 0	It6	0.142 (0.880)	0.084 2
It7	1.602* (2.393)	0.053 0						

Remarks: Values reported are the coefficients in the regression of equation (5.11). The t-statistics, displayed in parentheses beneath the estimates, relate to the two-sided test for $m = 0$. Preliminary regression include the number of lagged differences of the variable noted in the table in order to correct for autocorrelation. Additionally, the theoretically derived value μ is given. The results refer to OLS-regression using quarterly data deflated by PPI.

* indicates significant values at 5%-level.

A.5.3 Testing for unit roots in levels

	Model I					
	$\hat{\rho}$		$\hat{\mu}$		$\hat{\beta}$	
	ADF	PP	ADF	PP	ADF	PP
Indexes						
In1	0.749• (−3.212)	0.855 (−2.772)	0.222 (1.949)	0.185 (2.194)	0.024• (3.139)	0.014• (2.751)
In2	0.779 (−2.969)	0.864 (−2.668)	0.421 (2.371)	0.307 (2.315)	0.019• (2.786)	0.012• (2.528)
In3	0.797 (−2.887)	0.876 (−2.587)	0.175 (2.332)	0.128 (2.295)	0.008• (2.673)	0.005• (2.428)
Branch-Indexes						
Se1	0.783 (−2.932)	0.859 (−2.679)	0.190 (2.461)	0.142 (2.358)	0.007• (2.780)	0.005• (2.559)
Se2	0.838 (−2.375)	0.890 (−2.347)	0.181 (1.929)	0.144 (2.090)	0.005 (1.857)	0.004 (1.830)
Se3	0.857 (−2.467)	0.911 (−2.254)	0.128 (2.159)	0.096 (2.140)	0.003 (2.078)	0.002 (1.890)
Se4	0.872 (−2.414)	0.922 (−2.172)	0.077 (1.728)	0.047 (1.659)	0.004• (2.690)	0.003• (2.592)
Se5	0.845 (−2.291)	0.875 (−2.373)	0.141 (1.883)	0.128 (2.049)	0.005 (2.355)	0.004• (2.413)
Se6	0.774• (−3.359)	0.875 (−2.574)	0.241• (3.022)	0.142 (2.387)	0.002 (2.135)	0.001 (1.452)
Se7	0.856 (−2.161)	0.895 (−2.225)	0.201 (1.449)	0.172 (1.771)	0.007 (1.544)	0.005 (1.596)
Se8	0.810 (−3.033)	0.898 (−2.362)	0.156 (2.096)	0.109 (1.966)	0.006• (2.836)	0.003 (2.030)
Se9	0.780• (−3.253)	0.883 (−2.660)	0.200 (2.496)	0.113 (2.145)	0.007• (2.822)	0.004 (2.376)
Se10	0.819 (−2.666)	0.900 (−2.308)	0.106 (0.769)	0.141 (1.784)	0.017 (2.261)	0.009 (1.945)
Stocks						
St1	0.790 (−2.587)	0.863 (−2.345)	0.349 (1.988)	0.228 (1.788)	0.001 (0.371)	0.000 (0.348)
St2	0.856 (−2.128)	0.880 (−2.269)	0.070 (1.427)	0.066 (1.680)	0.005 (2.232)	0.004• (2.396)
St3	0.843 (−2.241)	0.875 (−2.369)	0.048 (1.044)	0.053 (1.459)	0.007• (2.393)	0.006• (2.513)
St4	0.666 (−2.851)	0.816 (−2.693)	−0.009 (−0.072)	0.028 (0.596)	0.030• (3.059)	0.018• (2.845)
St5	0.820• (−3.670)	0.918 (−2.133)	−0.020 (−0.106)	0.082 (1.060)	0.025• (3.000)	0.011 (1.781)
St6	0.806• (−3.133)	0.882 (−2.445)	0.099 (0.977)	0.111 (1.319)	0.016• (3.205)	0.010• (2.501)
St7	0.809 (−2.613)	0.906 (−2.133)	0.043 (0.693)	0.006 (0.744)	0.008• (2.947)	0.005• (2.550)
Dividends						
Div1	0.817 (−1.372)	0.867 (−1.763)	0.033 (1.052)	0.027 (1.606)	0.000 (0.748)	0.000 (0.465)
Div2	0.732 (−1.422)	0.791 (−2.007)	0,069 (1.072)	0,048 (1.540)	−0.229 (−1.133)	−0.167 (−1.577)
Div3	0.825 (−1.341)	0.850 (−1.524)	0.006 (0.131)	0.031 (0.934)	0.000 (0.130)	−0.001 (−0.841)

Remarks: Values reported are the coefficients in the regression of equation (5.8) to equation (5.10). The t-statistics reported in parentheses beneath the estimates relate to the tests for $\rho = 1$, $\beta = 0$ or $\mu = 0$. ADF-statistics include lags of differenced variable of order four. PP-statistics are corrected for autocorrelation and heteroskedasticity. Critical values are tabulated in appendix B.2. The results refer to OLS-regression using quarterly or annual data deflated by PPI. • indicates significant estimates at least at a 10 % level. o denotes processes where simple DF-statistics instead of ADF-statistics are reported because of lack of autocorrelation.

	Model I					
	$\hat{\rho}$		$\hat{\mu}$		$\hat{\beta}$	
	ADF	PP	ADF	PP	ADF	PP
Div4o	0.897 (−1.297)	− − (−1.313)	0.008 (0.355)	− − (0.298)	−0.000 (−0.047)	− − (0.013)
Div5o	0.725 (−2.510)	− − (−2.601)	0.045 (2.457)	− − (2.556)	−0.001 (−1.047)	− − (−1.135)
Div6	0.839 (−1.381)	0.892 (−1.343)	0.021 (0.474)	0.021 (0.829)	−0.000 (−0.183)	−0.000 (−0.642)
Div7	0.952 (−0.341)	0.901 (−0.992)	−0.025 (−0.458)	0.017 (0.466)	0.001 (0.622)	−0.000 (−0.361)
Div8	0.577 (−1.697)	0.669 (−2.543)	0.085 (1.254)	0.073 (2.186)	−0.002 (−1.170)	−0.002 (−2.091)
National accounting data						
Na1	0.955 (−2.226)	0.973 (−1.380)	0.152 (2.154)	0.102 (1.411)	0.003* (2.819)	0.002 (2.272)
Na2	0.885 (−2.566)	0.919 (−2.035)	0.069 (2.433)	0.055 (1.973)	0.002* (2.721)	0.001 (2.311)
Na3	0.949 (−2.384)	0.970 (−1.454)	0.161 (2.310)	0.106 (1.495)	0.003* (2.958)	0.002 (2.280)
Na4	0.918 (−2.323)	0.946 (−1.714)	0.130 (2.286)	0.092 (1.745)	0.001 (2.482)	0.001 (1.942)
Na5	0.945 (−2.386)	0.963 (−1.619)	0.028 (2.280)	0.020 (1.593)	0.000 (2.657)	0.000 (2.237)
Na6	0.990 (−0.838)	1.007 (0.287)	0.007 (0.804)	−0.002 (−0.033)	0.000 (2.038)	0.000 (1.578)
Proxies						
Pi1	0.928 (−2.001)	0.951 (−1.890)	6.421 (2.019)	4.735 (1.947)	0.035 (1.949)	0.022 (1.981)
Pi2o	0.737 (−2.960)	− −* (−3.197)	22.882* (3.088)	− −* (3.321)	0.122* (2.560)	− −* (2.887)
Pi3	0.921 (−2.081)	0.951 (−1.902)	6.927 (2.105)	4.715 (1.970)	0.040 (1.995)	0.022 (1.940)
Pi4	0.891 (−2.091)	0.919 (−2.250)	9.236 (2.114)	7.127 (2.283)	0.062 (2.284)	0.048* (2.645)
Pi5	0.904 (−2.133)	0.958 (−1.735)	8.032 (2.226)	4.226 (1.882)	0.058 (1.813)	0.017 (1.488)
Pi6	0.899 (−2.918)	0.955 (−1.787)	9.291* (2.905)	4.721 (1.856)	0.036 (2.302)	0.005 (1.068)
Pi7	0.851 (−1.491)	0.808 (−2.651)	12.832 (1.652)	16.249* (2.769)	0.083 (1.072)	0.111 (2.031)
Pi8	0.945 (−2.396)	0.969 (−1.657)	5.114 (2.323)	2.994 (1.653)	0.022 (2.266)	0.012 (1.612)
Pi9	0.957 (−0.862)	0.894 (−1.713)	3.281 (0.618)	10.116 (1.435)	0.060 (2.035)	0.076 (2.320)
Io1	0.919 (−2.062)	0.948 (−2.029)	6.834 (2.101)	4.854 (2.107)	0.048 (1.949)	0.029 (2.047)
Io2	0.863 (−2.429)	0.911 (−2.502)	10.938 (2.502)	7.644 (2.594)	0.091 (2.428)	0.058* (2.719)
Io3	0.912 (−2.159)	0.951 (−1.984)	7.064 (2.214)	4.630 (2.111)	0.059 (1.932)	0.026 (1.799)
Io4	0.883 (−2.434)	0.935 (−2.163)	10.329 (2.437)	6.210 (2.216)	0.040 (1.574)	0.016 (1.388)
Io5	0.922 (−1.258)	0.911 (−2.116)	6.564 (1.485)	7.412 (2.249)	0.039 (0.702)	0.053 (1.633)
Io6	0.933 (−1.870)	0.955 (−1.700)	6.189 (1.879)	4.415 (1.726)	0.033 (1.682)	0.022 (1.663)

Remarks: Values reported are the coefficients in the regression of equation (5.8) to equation (5.10). The t-statistics reported in parentheses beneath the estimates relate to the tests for $\rho = 1$, $\beta = 0$ or $\mu = 0$. ADF-statistics include lags of differenced variable of order four. PP-statistics are corrected for autocorrelation and heteroskedasticity. Critical values are tabulated in appendix B.2. The results refer to OLS-regression using quarterly or annual data deflated by PPI. • indicates significant estimates at least at a 10 % level. o denotes processes where simple DF-statistics instead of ADF-statistics are reported because of lack of autocorrelation.

	Model I					
	$\hat{\rho}$		$\hat{\mu}$		$\hat{\beta}$	
	ADF	PP	ADF	PP	ADF	PP
Io7	0.946	0.959	4.374	3.458	0.057	0.051
	(−1.843)	(−1.504)	(1.431)	(1.254)	(2.308)	(1.864)
It1	0.910	0.944	7.653	5.200	0.053	0.030
	(−2.128)	(−1.919)	(2.178)	(2.001)	(1.990)	(1.926)
It2	0.871	0.918	10.361	7.098	0.085	0.054•
	(−2.442)	(−2.438)	(2.494)	(2.517)	(2.510)	(2.724)
It3	0.892	0.937	8.770	5.744	0.072	0.036
	(−1.968)	(−1.772)	(2.074)	(1.907)	(1.698)	(1.573)
It4	0.907	0.917	8.460	7.840	0.029	0.019
	(−1.729)	(−1.767)	(1.816)	(1.850)	(1.051)	(0.994)
It5	0.826	0.819	14.047	15.149	0.117	0.111
	(−1.661)	(−2.537)	(1.789)	(2.680)	(1.297)	(1.896)
It6	0.941	0.969	5.364	3.016	0.028	0.014
	(−2.322)	(−1.644)	(2.292)	(1.670)	(2.176)	(1.650)
It7o	0.948	−−	4.031	−−	0.111	−−
	(−1.110)	(−1.027)	(1.054)	(0.979)	(1.562)	(1.527)

Remarks: Values reported are the coefficients in the regression of equation (5.8) to equation (5.10). The t-statistics reported in parentheses beneath the estimates relate to the tests for $\rho = 1$, $\beta = 0$ or $\mu = 0$. ADF-statistics include lags of differenced variable of order four. PP-statistics are corrected for autocorrelation and heteroskedasticity. Critical values are tabulated in appendix B.2. The results refer to OLS-regression using quarterly or annual data deflated by PPI. • indicates significant estimates at least at a 10 % level. o denotes processes where simple DF-statistics instead of ADF-statistics are reported because of lack of autocorrelation.

	Model II				Model III	
	$\bar{\rho}$		$\bar{\mu}$		$\breve{\rho}$	
	ADF	PP	ADF	PP	ADF	PP
	Indexes					
In1	0.981	0.989	0.140	0.120	1.010	1.013
	(−0.694)	(−0.665)	(1.172)	(1.182)	(0.879)	(1.007)
In2	0.967	0.977	0.213	0.174	1.006	1.008
	(−0.994)	(−1.000)	(1.248)	(1.264)	(0.534)	(0.558)
In3	0.967	0.977	0.093	0.075	1.005	1.008
	(−1.045)	(−1.029)	(1.285)	(1.284)	(0.483)	(0.527)
	Branch-Indexes					
Se1	0.969	0.977	0.083	0.071	1.006	1.008
	(−0.924)	(−0.967)	(1.170)	(1.214)	(0.591)	(0.589)
Se2	0.942	0.954	0.148	0.124	1.001	1.003
	(−1.458)	(−1.493)	(1.566)	(1.601)	(0.064)	(0.046)
Se3	0.961	0.969	0.073	0.065	0.999	1.003
	(−1.297)	(−1.328)	(1.342)	(1.415)	(−0.046)	(0.084)
Se4	1.001	1.013	0.018	0.003	1.011•	1.014•
	(0.030)	(0.101)	(0.436)	(0.411)	(1.382)	(1.495)
Se5	0.987	0.987	0.053	0.053	1.011	1.010
	(−0.410)	(−0.541)	(0.785)	(0.909)	(1.109)	(1.083)
Se6	0.859	0.908	0.197•	0.131	0.995	0.998
	(−2.521)	(−2.195)	(2.481)	(2.172)	(−0.430)	(−0.309)
Se7	0.934	0.948	0.227	0.191	1.000	1.002
	(−1.500)	(−1.543)	(1.628)	(1.675)	(−0.000)	(−0.010)
Se8	0.958	0.965	0.089	0.080	0.999	1.002
	(−1.139)	(−1.325)	(1.194)	(1.393)	(−0.001)	(0.009)

Remarks: Values reported are the coefficients in the regression of equation (5.8) to equation (5.10). The t-statistics reported in parentheses beneath the estimates relate to the tests for $\rho = 1$, $\beta = 0$ or $\mu = 0$. ADF-statistics include lags of differenced variable of order four. PP-statistics are corrected for autocorrelation and heteroskedasticity. Critical values are tabulated in appendix B.2. The results refer to OLS-regression using quarterly or annual data deflated by PPI. • indicates significant estimates at least at a 10 % level. o denotes processes where simple DF-statistics instead of ADF-statistics are reported because of lack of autocorrelation.

	Model II				Model III	
	$\tilde{\rho}$		$\tilde{\mu}$		$\tilde{\rho}$	
	ADF	PP	ADF	PP	ADF	PP
Se9	0.939	0.966	0.133	0.082	1.001	1.003
	(−1.540)	(−1.385)	(1.641)	(1.466)	(0.067)	(0.016)
Se10	0.957	0.970	0.219	0.190	1.001	1.007
	(−1.401)	(−1.299)	(1.650)	(1.643)	(0.071)	(0.234)
Stocks						
St1	0.779*	0.856	0.389*	0.257*	0.989	0.989
	(−2.916)	(−2.584)	(2.819)	(2.451)	(−0.766)	(−0.851)
St2	0.993	0.996	0.041	0.038	1.016*	1.016*
	(−0.227)	(−0.234)	(0.835)	(0.858)	(1.390)	(1.478)
St3	0.999	0.999	0.039	0.038	1.018*	1.017*
	(−0.030)	(−0.125)	(0.810)	(0.889)	(1.558)	(1.616)
St4	1.005	0.997	0.090	0.103	1.030*	1.022*
	(0.136)	(−0.052)	(0.691)	(0.906)	(1.493)	(1.553)
St5	0.942	0.972	0.314	0.217	0.988	1.004
	(−1.980)	(−1.197)	(1.878)	(1.391)	(−0.732)	(−0.065)
St6	0.984	0.987	0.091	0.087	1.007	1.008
	(−0.540)	(−0.595)	(0.838)	(0.926)	(0.550)	(0.642)
St7	0.994	1.010	0.039	0.022	1.016	1.020
	(−0.146)	(−0.127)	(0.582)	(0.643)	(0.997)	(1.047)
Dividends						
Div1	0.790	0.864	0.045	0.030	0.999	0.999
	(−1.650)	(−1.828)	(1.652)	(1.834)	(−0.071)	(−0.073)
Div2	0.937	0.942	−0.003	−0.002	0.922*	0.935*
	(−1.189)	(−1.456)	(−0.390)	(−0.293)	(−2.106)	(−2.598)
Div3	0.811	0.919	0.012	0.005	0.875*	0.943*
	(−2.498)	(−1.399)	(1.051)	(0.328)	(−2.782)	(−2.051)
Div4o	0.899	−−	0.007	−−	0.937*	−−*
	(−1.919)	(−1.990)	(0.847)	(0.849)	(−2.211)	(−2.282)
Div5o	0.732	−−*	0.032*	−−*	0.963	−−
	(−2.449)	(−2.597)	(2.359)	(2.486)	(−0.711)	(−0.766)
Div6	0.857	0.929	0.013	0.006	0.918*	0.954*
	(−2.403)	(−1.395)	(1.241)	(0.604)	(−2.407)	(−1.644)
Div7	0.872	0.936	0.009	0.004	0.912*	0.955*
	(−2.129)	(−1.384)	(0.837)	(0.418)	(−2.436)	(−1.919)
Div8	0.849	0.908	0.007	0.006	0.895*	0.942*
	(−1.678)	(−1.230)	(0.599)	(0.319)	(−2.210)	(−1.860)
National accounting data						
Na1	1.008	1.018	−0.019	−0.052	1.004*	1.009*
	(0.993)	(1.912)	(−0.496)	(0.965)	(2.194)	(5.019)
Na2	0.999	1.004	0.008	0.006	1.006*	1.009*
	(−0.047)	(0.151)	(0.455)	(0.353)	(1.543)	(2.104)
Na3	1.008	1.017	−1.836	−4.397	1.004*	1.008*
	(0.998)	(1.750)	(−0.501)	(−0.851)	(2.221)	(4.883)
Na4	0.998	1.001	0.010	0.007	1.003*	1.004*
	(−0.150)	(−0.166)	(0.322)	(0.390)	(1.344)	(1.854)
Na5	1.000	1.013	0.002	−0.001	1.003*	1.007*
	(0.007)	(0.684)	(0.249)	(−0.205)	(1.331)	(2.815)
Na6	1.010	1.026*	−0.005	−0.014	1.006*	1.017*
	(1.493)	(6.077)	(−0.706)	(−2.055)	(2.122)	(11.455)
Proxies						
Pi1	0.991	0.990	1.136	1.384	1.002	1.003*
	(−0.556)	(−0.843)	(0.668)	(0.966)	(1.243)	(1.339)
Pi2o	0.957	−−	4.814	−−*	1.004*	−−*
	(−1.854)	(−1.956)	(2.050)	(2.187)	(1.776)	(2.186)

Remarks: Values reported are the coefficients in the regression of equation (5.8) to equation (5.10). The t-statistics reported in parentheses beneath the estimates relate to the tests for $\rho = 1$, $\beta = 0$ or $\mu = 0$. ADF-statistics include lags of differenced variable of order four. PP-statistics are corrected for autocorrelation and heteroskedasticity. Critical values are tabulated in appendix B.2. The results refer to OLS-regression using quarterly or annual data deflated by PPI. • indicates significant estimates at least at a 10 % level. o denotes processes where simple DF-statistics instead of ADF-statistics are reported because of lack of autocorrelation.

	Model II				Model III	
	$\bar{\rho}$		$\bar{\mu}$		$\breve{\rho}$	
	ADF	PP	ADF	PP	ADF	PP
Pi3	0.989 (−0.635)	0.989 (−0.910)	1.322 (0.752)	1.583 (1.040)	1.002 (1.186)	1.004• (1.300)
Pi4	1.005 (0.347)	1.007 (−0.067)	−0.142 (−0.092)	−0.151 (0.286)	1.004• (2.545)	1.005• (2.220)
Pi5	0.979 (−1.154)	0.979 (−1.237)	2.457 (1.276)	2.635 (1.386)	1.002 (0.907)	1.004 (1.073)
Pi6	0.959 (−1.734)	0.961 (−1.639)	4.332 (1.767)	4.213 (1.671)	1.001 (0.353)	1.001 (0.289)
Pi7	0.948 (−1.219)	0.931 (−1.632)	5.991 (1.350)	7.554 (1.762)	1.005 (0.864)	1.004 (0.827)
Pi8	0.984 (−1.034)	0.988 (−1.102)	1.813 (1.061)	1.418 (1.135)	1.000 (0.349)	1.001 (0.410)
Pi9	1.008 (0.191)	0.959 (−0.441)	−0.476 (−0.093)	5.053 (0.503)	1.004 (1.092)	1.002 (0.700)
Io1	0.988 (−0.691)	0.989 (−0.957)	1.539 (0.841)	1.682 (1.112)	1.003 (1.273)	1.005 (1.282)
Io2	0.996 (−0.287)	0.996 (−0.548)	0.868 (0.603)	1.068 (0.847)	1.004• (2.379)	1.007• (2.366)
Io3	0.982 (−0.942)	0.981 (−1.199)	2.105 (1.085)	2.462 (1.360)	1.003 (0.980)	1.005 (1.063)
Io4	0.932 (−1.833)	0.952 (−1.847)	6.956 (1.878)	5.114 (1.902)	1.001 (0.382)	1.003 (0.430)
Io5	0.961 (−1.358)	0.959 (−1.488)	4.241 (1.452)	4.528 (1.597)	1.002 (0.446)	1.004 (0.492)
Io6	0.983 (−0.842)	0.987 (−0.901)	2.069 (0.925)	1.694 (0.984)	1.002 (0.858)	1.003 (0.857)
Io7	0.982 (−0.702)	0.987 (−0.881)	2.267 (0.749)	1.851 (0.939)	1.001 (0.355)	1.002 (0.391)
It1	0.988 (−0.737)	0.989 (−0.845)	1.515 (0.878)	1.614 (1.006)	1.002 (1.185)	1.004• (1.373)
It2	1.000 (−0.008)	1.000 (−0.369)	0.431 (0.326)	0.645 (0.669)	1.004• (2.485)	1.007• (2.475)
It3	0.979 (−1.058)	0.980 (−1.059)	2.464 (1.198)	2.514 (1.236)	1.003 (0.927)	1.004 (1.187)
It4	0.952 (−1.483)	0.945 (−1.544)	4.952 (1.521)	5.746 (1.584)	1.001 (0.365)	1.001 (0.360)
It5	0.950 (−1.168)	0.931 (−1.708)	5.614 (1.266)	7.574 (1.819)	1.004 (0.550)	1.003 (0.597)
It6	0.987 (−0.903)	0.991 (−0.930)	1.532 (0.963)	1.183 (1.000)	1.001 (0.633)	1.002 (0.734)
It7o	1.013 (0.550)	−− (0.833)	0.058 (0.020)	−− (−0.169)	1.013• (2.462)	−−• (2.871)

Remarks: Values reported are the coefficients in the regression of equation (5.8) to equation (5.10). The t-statistics reported in parentheses beneath the estimates relate to the tests for $\rho = 1$, $\beta = 0$ or $\mu = 0$. ADF-statistics include lags of differenced variable of order four. PP-statistics are corrected for autocorrelation and heteroskedasticity. Critical values are tabulated in appendix B.2. The results refer to OLS-regression using quarterly or annual data deflated by PPI. • indicates significant estimates at least at a 10 % level. o denotes processes where simple DF-statistics instead of ADF-statistics are reported because of lack of autocorrelation.

A.5.4 Testing for unit roots in logs

	Model I					
	$\hat{\rho}$		$\hat{\mu}$		$\hat{\beta}$	
	ADF	PP	ADF	PP	ADF	PP
Dividends						
Log Div1	0.820	0.862	0.518	0.413	0.001	0.001
	(−1.314)	(−1.788)	(1.208)	(1.742)	(0.761)	(0.776)
Log Div2	−0.163*	−0.527*	5.072*	1.895*	−0.144*	−0.059*
	(−3.793)	(−3.240)	(3.711)	(3.052)	(−3.845)	(−3.176)
Log Div3	0.822	0.849	0.428	0.447	−0.003	−0.006
	(−1.270)	(−1.659)	(0.800)	(1.400)	(−0.314)	(−1.098)
Log Div4	0.772	0.856	0.384	0.351	0.004	−0.002
	(−1.731)	(−1.450)	(0.848)	(1.001)	(0.454)	(−0.299)
Log Div5o	0.769	−−	0.642*	−−*	−0.004	−−
	(−2.464)	(−2.549)	(2.933)	(3.098)	(−0.648)	(−0.870)
Log Div6	0.878	0.947	0.271	0.100	−0.000	−0.000
	(−1.124)	(−0.930)	(0.681)	(0.649)	(−0.037)	(−0.139)
Log Div7	0.912	0.953	0.158	0.073	0.001	0.001
	(−0.614)	(−0.639)	(0.273)	(0.334)	(0.112)	(0.058)
Log Div8	0.596	0.725	1.340	0.881	−0.024	−0.016
	(−1.881)	(−2.438)	(1.689)	(2.280)	(−1.559)	(−2.105)
National accounting data						
Log Na1	0.943	0.967	0.338	0.200	0.001*	0.000
	(−2.478)	(−1.548)	(2.481)	(1.562)	(2.824)	(2.438)
Log Na2o	0.926	−−	0.319	−−	0.001	−−
	(−1.864)	(−1.990)	(1.873)	(1.998)	(2.075)	(2.295)
Log Na3	0.937	0.964	0.368	0.211	0.001*	0.000
	(−2.608)	(−1.583)	(2.610)	(1.597)	(2.955)	(2.417)
Log Na4	0.917	0.948	0.424	0.265	0.001	0.000
	(−2.367)	(−1.687)	(2.364)	(1.695)	(2.506)	(2.318)
Log Na5	0.937	0.961	0.252	0.158	0.001*	0.000
	(−2.555)	(−1.658)	(2.546)	(1.663)	(2.792)	(2.402)
Log Na6	0.963	0.984	0.167	0.081	0.001	0.000
	(−1.804)	(−0.996)	(1.861)	(1.078)	(2.223)	(1.604)
Proxies						
Log Pi1	0.927	0.950	0.327	0.229	0.000	0.000*
	(−1.959)	(−1.881)	(1.965)	(1.894)	(1.900)	(2.998)
Log Pi2o	0.705*	−−*	1.321*	−−*	0.001*	−−*
	(−3.342)	(−3.697)	(3.370)	(3.722)	(2.866)	(3.970)
Log Pi3	0.920	0.949	0.358	0.232	0.000	0.000*
	(−2.015)	(−1.890)	(2.023)	(1.906)	(1.921)	(2.880)
Log Pi4	0.890	0.909	0.492	0.408	0.001	0.000*
	(−2.107)	(−2.376)	(2.114)	(2.385)	(2.235)	(3.731)
Log Pi5	0.900	0.952	0.446	0.222	0.001	0.000
	(−2.031)	(−1.724)	(2.053)	(1.755)	(1.699)	(2.094)
Log Pi6	0.898	0.949	0.464*	0.235	0.000	0.000
	(−2.863)	(−1.822)	(2.865)	(1.837)	(2.253)	(2.137)
Log Pi7	0.825	0.745	0.781	1.129*	0.001	0.001
	(−1.519)	(−3.052)	(1.551)	(3.081)	(1.097)	(2.304)
Log Pi8	0.945	0.969	0.249	0.141	0.000	0.000*
	(−2.404)	(−1.649)	(2.392)	(1.649)	(2.294)	(2.892)
Log Pi9	0.952	0.887	0.213	0.518	0.001	0.001
	(−0.947)	(−1.819)	(0.904)	(1.772)	(1.981)	(1.671)
Log Io1	0.922	0.947	0.349	0.241	0.000	0.000*
	(−1.976)	(−2.039)	(1.989)	(2.058)	(1.838)	(2.976)

Remarks: Values reported are the coefficients in the regression of equation (5.8) to equation (5.10). The t-statistics reported in parentheses beneath the estimates relate to the tests for $\rho = 1$, $\beta = 0$ or $\mu = 0$. ADF-statistics include lags of differenced variable of order four. PP-statistics are corrected for autocorrelation and heteroskedasticity. Critical values are tabulated in appendix B.2. The results refer to OLS–regression using quarterly or annual data deflated by PPI. * indicates significant estimates at least at a 10 % level. o denotes processes where simple DF–statistics instead of ADF–statistics are reported because of lack of autocorrelation.

	Model I					
	$\hat{\rho}$		$\hat{\mu}$		$\hat{\beta}$	
	ADF	PP	ADF	PP	ADF	PP
Log Io2	0.868 (−2.362)	0.902 (−2.618)	0.583 (2.380)	0.438 (2.641)	0.001 (2.287)	0.001• (3.690)
Log Io3	0.913 (−2.082)	0.949 (−2.033)	0.385 (2.099)	0.233 (2.032)	0.001 (1.846)	0.000• (2.580)
Log Io4	0.881 (−2.444)	0.933 (−2.201)	0.534 (2.453)	0.307 (2.217)	0.000 (1.571)	0.000• (2.465)
Log Io5	0.916 (−1.272)	0.899 (−2.146)	0.373 (1.316)	0.444 (2.179)	0.000 (0.758)	0.001 (2.029)
Log Io6	0.935 (−1.843)	0.955 (−1.690)	0.297 (1.847)	0.208 (1.696)	0.000 (1.669)	0.000• (2.421)
Log Io7	0.945 (−1.834)	0.957 (−1.582)	0.245 (1.765)	0.192 (1.539)	0.000 (2.308)	0.000 (2.258)
Log Is1	0.906 (−2.061)	0.939 (−1.913)	0.418 (2.074)	0.277 (1.931)	0.001 (1.918)	0.000• (2.595)
Log Is2	0.871 (−2.411)	0.905 (−2.580)	0.567 (2.426)	0.423 (2.600)	0.001 (2.396)	0.001• (3.765)
Log Is3	0.887 (−1.853)	0.922 (−1.790)	0.501 (1.876)	0.349 (1.817)	0.001 (1.587)	0.000 (1.884)
Log Is4	0.907 (−1.656)	0.906 (−1.850)	0.418 (1.676)	0.428 (1.868)	0.000 (0.988)	0.000 (1.325)
Log Is5	0.780 (−1.774)	0.747 (−3.020)	0.968 (1.799)	1.118• (3.053)	0.001 (1.438)	0.002 (2.362)
Log Is6	0.942 (−2.304)	0.969 (−1.633)	0.262 (2.300)	0.140 (1.638)	0.000 (2.179)	0.000• (2.704)
Log Is7o	0.885 (−1.892)	— (−1.832)	0.512 (1.911)	— (1.853)	0.001 (1.926)	— (1.781)

Remarks: Values reported are the coefficients in the regression of equation (5.8) to equation (5.10). The t-statistics reported in parentheses beneath the estimates relate to the tests for $\rho = 1$, $\beta = 0$ or $\mu = 0$. ADF-statistics include lags of differenced variable of order four. PP-statistics are corrected for autocorrelation and heteroskedasticity. Critical values are tabulated in appendix B.2. The results refer to OLS-regression using quarterly or annual data deflated by PPI. • indicates significant estimates at least at a 10 % level. o denotes processes where simple DF-statistics instead of ADF-statistics are reported because of lack of autocorrelation.

	Model II				Model III	
	$\tilde{\rho}$		$\tilde{\mu}$		$\breve{\rho}$	
	ADF	PP	ADF	PP	ADF	PP
Dividends						
Log Div1	0.792 (−1.596)	0.858 (−1.858)	0.634 (1.599)	0.436 (1.863)	1.000 (0.044)	1.000 (0.066)
Log Div2	0.978 (−0.230)	0.948 (−0.605)	−0.132 (−0.560)	−0.007 (−0.322)	0.930 (−1.525)	0.946 (−1.605)
Log Div3	0.859 (−1.864)	0.929 (−1.214)	0.271 (1.459)	0.131 (0.813)	0.966• (−1.867)	0.979• (−1.807)
Log Div4	0.723• (−3.037)	0.883 (−1.874)	0.565• (2.723)	0.246 (1.548)	0.967• (−1.665)	0.980• (−1.554)
Log Div5o	0.752• (−2.774)	—• (−2.858)	0.603• (2.892)	—• (2.985)	1.000 (−0.003)	— (−0.111)
Log Div6	0.882 (−2.286)	0.941 (−1.425)	0.266 (2.001)	0.123 (1.149)	0.983 (−1.403)	0.986 (−1.375)
Log Div7	0.898 (−1.924)	0.944 (−1.360)	0.221 (1.654)	0.112 (1.042)	0.983 (−1.218)	0.985 (−1.418)

Remarks: Values reported are the coefficients in the regression of equation (5.8) to equation (5.10). The t-statistics reported in parentheses beneath the estimates relate to the tests for $\rho = 1$, $\beta = 0$ or $\mu = 0$. ADF-statistics include lags of differenced variable of order four. PP-statistics are corrected for autocorrelation and heteroskedasticity. Critical values are tabulated in appendix B.2. The results refer to OLS-regression using quarterly or annual data deflated by PPI. • indicates significant estimates at least at a 10 % level. o denotes processes where simple DF-statistics instead of ADF-statistics are reported because of lack of autocorrelation.

	Model II				Model III	
	$\tilde{\rho}$		$\tilde{\mu}$		$\breve{\rho}$	
	ADF	PP	ADF	PP	ADF	PP
Log Div8	0.909 (−1.153)	0.935 (−0.937)	0.134 (0.735)	0.096 (0.511)	0.965* (−1.614)	0.974* (−1.625)
National accounting data						
Log Na1	1.005 (0.622)	1.013 (1.190)	−0.026 (−0.551)	−0.074 (−1.060)	1.001* (2.005)	1.001* (4.317)
Log Na2o	1.001 (0.053)	– – (−0.089)	0.004 (0.055)	– – (0.185)	1.002* (2.083)	– –* (1.855)
Log Na3	1.005 (0.633)	1.012 (1.080)	−0.027 (−0.562)	−0.068 (−0.954)	1.001* (2.023)	1.001* (4.215)
Log Na4	1.000 (−0.214)	0.997 (−0.250)	0.020 (0.244)	0.007 (0.291)	1.000 (1.282)	1.001* (1.813)
Log Na5	1.000 (−0.020)	1.011 (0.514)	0.003 (0.075)	−0.039 (−0.411)	1.001* (1.333)	1.001* (2.653)
Log Na6	1.008 (1.661)	1.011* (2.715)	−0.028 (−1.360)	−0.040 (−1.952)	1.001* (2.634)	1.003* (10.725)
Proxies						
Log Pi1	0.991 (−0.562)	0.989 (−0.875)	0.045 (0.586)	0.055 (0.902)	1.000* (1.307)	1.001* (1.446)
Log Pi2o	0.950 (−2.142)	– – (−2.246)	0.237* (2.188)	– –* (2.298)	1.001* (1.997)	– –* (2.330)
Log Pi3	0.989 (−0.636)	0.987 (−0.940)	0.053 (0.663)	0.064 (0.970)	1.000 (1.272)	1.001* (1.434)
Log Pi4	1.002 (0.129)	1.000 (−0.345)	−0.005 (−0.075)	0.004 (0.004)	1.001* (2.473)	1.001* (2.106)
Log Pi5	0.977 (−1.173)	0.976 (−1.279)	0.110 (1.203)	0.110 (1.316)	1.001 (1.098)	1.001* (1.330)
Log Pi6	0.959 (−1.712)	0.958 (−1.651)	0.193 (1.720)	0.195 (1.659)	1.000 (0.466)	1.000 (0.427)
Log Pi7	0.940 (−1.245)	0.913 (−1.771)	0.286 (1.277)	0.405 (1.804)	1.002 (1.075)	1.001 (1.088)
Log Pi8	0.984 (−1.023)	0.989 (−1.074)	0.077 (1.029)	0.054 (1.081)	1.000 (0.396)	1.000 (0.484)
Log Pi9	1.002 (−0.036)	0.950 (−0.629)	−0.004 (−0.018)	0.240 (0.641)	1.001 (0.996)	1.000 (0.637)
Log Io1	0.987 (−0.722)	0.986 (−1.030)	0.061 (0.756)	0.069 (1.065)	1.001* (1.399)	1.001* (1.422)
Log Io2	0.992 (−0.569)	0.988 (−0.918)	0.042 (0.637)	0.063 (0.983)	1.001* (2.359)	1.001* (2.371)
Log Io3	0.982 (−0.944)	0.979 (−1.264)	0.084 (0.979)	0.102 (1.303)	1.001 (1.166)	1.001* (1.300)
Log Io4	0.931 (−1.849)	0.950 (−1.880)	0.320 (1.860)	0.233 (1.894)	1.000 (0.527)	1.001 (0.604)
Log Io5	0.961 (−1.334)	0.958 (−1.466)	0.184 (1.364)	0.198 (1.497)	1.001 (0.788)	1.001 (0.816)
Log Io6	0.983 (−0.812)	0.987 (−0.881)	0.079 (0.831)	0.062 (0.899)	1.000 (0.945)	1.001 (0.949)
Log Io7	0.981 (0.713)	0.985 (0.996)	0.091 (0.723)	0.076 (1.007)	1.000 (0.394)	1.000 (0.409)
Log Is1	0.987 (−0.739)	0.987 (−0.874)	0.062 (0.772)	0.065 (0.911)	1.001* (1.311)	1.001* (1.537)
Log Is2	0.996 (−0.338)	0.992 (−0.754)	0.025 (0.408)	0.045 (0.818)	1.001* (2.510)	1.001* (2.456)
Log Is3	0.978 (−1.050)	0.978 (−1.084)	0.105 (1.086)	0.107 (1.128)	1.001 (1.180)	1.001* (1.465)

Remarks: Values reported are the coefficients in the regression of equation (5.8) to equation (5.10). The t-statistics reported in parentheses beneath the estimates relate to the tests for $\rho = 1$, $\beta = 0$ or $\mu = 0$. ADF-statistics include lags of differenced variable of order four. PP-statistics are corrected for autocorrelation and heteroskedasticity. Critical values are tabulated in appendix B.2. The results refer to OLS-regression using quarterly or annual data deflated by PPI. * indicates significant estimates at least at a 10 % level. o denotes processes where simple DF-statistics instead of ADF-statistics are reported because of lack of autocorrelation.

	Model II				Model III	
	$\tilde{\rho}$		$\tilde{\mu}$		$\check{\rho}$	
	ADF	PP	ADF	PP	ADF	PP
Log Is4	0.952 (−1.458)	0.940 (−1.563)	0.223 (1.468)	0.277 (1.573)	1.000 (0.508)	1.000 (0.503)
Log Is5	0.945 (−1.146)	0.913 (−1.850)	0.263 (1.173)	0.406 (1.880)	1.001 (0.797)	1.001 (0.898)
Log Is6	0.987 (−0.868)	0.992 (−0.897)	0.061 (0.881)	0.040 (0.912)	1.000 (0.704)	1.000 (0.833)
Log Is7o	0.988 (−0.399)	−− (−0.119)	0.068 (0.487)	−− (0.227)	1.003• (1.819)	−−• (2.279)

Remarks: Values reported are the coefficients in the regression of equation (5.8) to equation (5.10). The t-statistics reported in parentheses beneath the estimates relate to the tests for $\rho = 1$, $\beta = 0$ or $\mu = 0$. ADF-statistics include lags of differenced variable of order four. PP-statistics are corrected for autocorrelation and heteroskedasticity. Critical values are tabulated in appendix B.2. The results refer to OLS-regression using quarterly or annual data deflated by PPI. • indicates significant estimates at least at a 10 % level. o denotes processes where simple DF-statistics instead of ADF-statistics are reported because of lack of autocorrelation.

Appendix B

Distributional Properties and Critical Values

B.1 Distributional properties of t–statistics

Auxiliary Regression $y_t = \hat{\rho} y_{t-1} + \hat{\mu} + \hat{\beta} t + v_t$	Data Generating Process $y_t = \rho y_{t-1} + \mu + \beta t + \epsilon_t$	Asymptotic Distr. for the t-statistic	Small sample Distr. for the t-statistic
$t_{\hat{\rho}}$ omitting trend and constant	$\rho < 1$ any μ, any β	Normal	
	$\rho = 1$ $\mu = 0$, $\beta = 0$	Brownian motion	Fuller (1976)
$t_{\hat{\rho}}$ omitting trend	$\rho < 1$ any μ, any β	Normal	
	$\rho = 1$ $\mu = 0$, $\beta = 0$	Brownian motion	Fuller (1976)
	$\rho = 1$ $\mu \neq 0$, $\beta = 0$	Normal	Hylleberg, Mizon (1989, 1995)
$t_{\hat{\mu}}$ omitting trend	$\rho < 1$ any μ, any β	Normal	
	$\rho = 1$ $\mu = 0$, $\beta = 0$	Brownian motion	Dickey, Fuller (1981)
	$\rho = 1$ $\mu \neq 0$, $\beta = 0$	Normal	Hylleberg, Haldrup (1989) tbl 2
$t_{\hat{\rho}}$	$\rho < 1$ any μ, any β	Normal	
	$\rho = 1$ any μ, $\beta = 0$	Brownian motion	Fuller (1976)
	$\rho = 1$ any μ, $\beta \neq 0$	Normal	
$t_{\hat{\mu}}$	$\rho < 1$ any μ, any β	Normal	
	$\rho = 1$ $\mu = 0$, $\beta = 0$	Brownian motion	Dickey, Fuller (1981) tbl 2
	$\rho = 1$ $\mu \neq 0$, $\beta = 0$	Brownian motion	Hylleberg, Haldrup (1989) tbl 3
	$\rho = 1$ any μ, $\beta \neq 0$	Normal	
$t_{\hat{\beta}}$	$\rho < 1$ $\mu \neq 0$, $\beta = 0$	Normal	
	$\rho = 1$ $\mu = 0$, $\beta = 0$	Brownian motion	Dickey, Fuller (1981) tbl 3
	$\rho = 1$ $\mu \neq 0$, $\beta = 0$	Brownian motion	Hylleberg, Haldrup (1989) tbl 4
	$\rho = 1$ any μ, $\beta \neq 0$	Normal	

Source: Hylleberg, Haldrup (1989), p.6
Notice that we do not distinguish between estimates of different models through using distinct signs above the coefficients as it is done in the text equation (5.8) to (5.10). The correspondence of the estimates is obvious.

B.2 Critical values of t–statistics

Model I estimated by OLS

$$z_t = \hat{\rho} z_{t-1} + \hat{\mu} + \hat{\beta} t + v_t$$

a. H_0 **: Random walk**

	$H_0 : z_t = z_{t-1} + \epsilon_t$	$H_1 : z_t = \rho z_{t-1} + \mu + \beta t + \epsilon_t$
t-Statistics for unit root		
$t_{\hat{\rho}}$, $Z(t_{\hat{\rho}})$	Fuller (1976), p.373	
t-Statistics for the constant		
$t_{\hat{\mu}}$, $Z(t_{\hat{\mu}})$	Dickey, Fuller (1981) Table II	
t-Statistics for the trend		
$t_{\hat{\beta}}$, $Z(t_{\hat{\beta}})$	Dickey, Fuller (1981), Table III	

	Critical values for T=50 [T=250]		
	$t_{\hat{\rho}}$, $Z(t_{\hat{\rho}})$	$t_{\hat{\mu}}$, $Z(t_{\hat{\mu}})$	$t_{\hat{\beta}}$, $Z(t_{\hat{\beta}})$
1%	-4.15 [-3.99]	3.87 [3.74]	3.60 [3.49]
5%	-3.50 [-3.43]	3.14 [3.09]	2.81 [2.79]
10%	-3.18 [-3.13]	2.75 [2.73]	2.38 [2.38]

b. H_0 **: Random walk with drift**

	$H_0 : z_t = z_{t-1} + \mu + \epsilon_t$	$H_1 : z_t = \rho z_{t-1} + \mu + \beta t + \epsilon_t$
t-Statistics for unit root		
$t_{\hat{\rho}}$	Fuller (1976), p.373; for critical values see above	
t-Statistics for the constant		
$t_{\hat{\mu},\ \mu=0.1}$	Haldrup, Hylleberg (1989), Table 3	
$t_{\hat{\mu},\ \mu=0.5}$	Haldrup, Hylleberg (1989), Table 3	
t-Statistics for the trend		
$t_{\hat{\beta},\ \mu=0.1}$	Haldrup, Hylleberg (1989), Table 4	
$t_{\hat{\beta},\ \mu=0.5}$	Haldrup, Hylleberg (1989), Table 4	

	Critical values for T=50 [T=250]			
	$t_{\hat{\mu},\ \mu=0.1}$	$t_{\hat{\mu},\ \mu=0.5}$	$t_{\hat{\beta},\ \mu=0.1}$	$t_{\hat{\beta},\ \mu=0.5}$
1%	4.13 [4.17]	4.49 [4.31]	3.82 [3.87]	4.10 [4.00]
5%	3.48 [3.63]	3.88 [3.79]	3.12 [3.28]	3.46 [3.44]
10%	3.13 [3.34]	3.58 [3.51]	2.74 [2.96]	3.14 [3.13]

Model II estimated by OLS

$$z_t = \tilde{\rho} z_{t-1} + \tilde{\mu} + v_t$$

a. H_0 **: Random walk**

	$H_0 : z_t = z_{t-1} + \epsilon_t$	$H_1 : z_t = \rho z_{t-1} + \mu + \epsilon_t$
t-Statistics for unit root		
$t_{\tilde{\rho}}$, $Z(t_{\tilde{\rho}})$	Fuller (1976), p.373	
t-Statistics for the constant		
$t_{\tilde{\mu}}$, $Z(t_{\tilde{\mu}})$	Dickey, Fuller (1981) Table I	
	Critical values for T=50 [T=250]	
	$t_{\tilde{\rho}}$, $Z(t_{\tilde{\rho}})$	$t_{\tilde{\mu}}$, $Z(t_{\tilde{\mu}})$
1%	-3.58 [-3.46]	3.28 [3.19]
5%	-2.93 [-2.88]	2.56 [2.53]
10%	-2.60 [-2.57]	2.18 [2.16]

b. H_0 **: Random walk with drift**

	$H_0 : z_t = z_{t-1} + \mu + \epsilon_t$		$H_1 : z_t = \rho z_{t-1} + \mu + \epsilon_t$	
t-Statistics for unit root				
$t_{\tilde{\rho},\ \mu=0.1}$	Hylleberg, Mizon (1989a), Table 1			
$t_{\tilde{\rho},\ \mu=0.5}$	Hylleberg, Mizon (1989a), Table 1			
t-Statistics for the constant				
$t_{\tilde{\mu},\ \mu=0.1}$	Haldrup, Hylleberg (1989), Table 2			
$t_{\tilde{\mu},\ \mu=0.5}$	Haldrup, Hylleberg (1989), Table 2			
	Critical values for T=50 [T=250]			
	$t_{\tilde{\rho},\ \mu=0.1}$	$t_{\tilde{\rho},\ \mu=0.5}$	$t_{\tilde{\mu},\ \mu=0.1}$	$t_{\tilde{\mu},\ \mu=0.5}$
1%	-3.35 [-3.27]	-2.57 [-2.63]	3.50 [3.59]	4.41 [6.18]
5%	-2.75 [-2.70]	-2.02 [-1.94]	2.81 [2.98]	3.69 [5.59]
10%	-2.47 [-2.38]	-1.72 [-1.61]	2.46 [2.68]	3.35 [5.28]

Model III estimated by OLS

$$z_t = \breve{\rho} z_{t-1} + v_t$$

a. H_0 : Random walk

	$H_0 : z_t = z_{t-1} + \epsilon_t$	$H_1 : z_t = \rho z_{t-1} + \epsilon_t$
t-Statistics for unit root		
$t_{\breve{\rho}}$, $Z(t_{\breve{\rho}})$	Fuller (1976), p.373	

	Critical values for T=50 [T=250]
	$t_{\breve{\rho}}$, $Z(t_{\breve{\rho}})$
1%	-2.62 [-2.58]
5%	-1.95 [-1.95]
10%	-1.61 [-1.62]

Bibliography

[1] AGENOR, P.-R. AND R. FLOOD (1994), Macroeconomic policy, speculative attacks, and balance of payments crisis, in: F. van der Ploeg (ed.), The Handbook of International Economics, Basil Blackwell, Cambridge, 224–250

[2] ALLAN, F. AND G. GORTON (1991), Rational finite bubbles, NBER Working Paper No. 3707

[3] ALLEN, F. AND G. GORTON (1993), Churning bubbles, Review of Economic Studies, 60, 813–83

[4] ARROW, K. (1982), Risk perception in psychology and economics, Economic Inquiry, 20, 1–9

[5] AZARIADIS, C. (1981), Self-fulfilling prophecies, Journal of Economic Theory, 25, 380–396

[6] AZARIADIS, C. AND R. GUESNERIE (1986), Sunspots and cycles, Review of Economic Studies

[7] BACHELIER, L. (1900), Theory of speculation, in: P. Cootner (1964), The random character of stock market prices, MIT University Press, Cambridge, 17–78

[8] BANERJEE, A. AND D. HENDRY (1992), Testing integration and cointegration: An overview, Oxford Bulletin of Economics and Statistics, 54, 225–255

[9] BEGG, D. (1982), The rational expectation revolution in macroeconomics, John-Hopkins University Press, Baltimore

[10] BERTOLA, G. (1994), Continuous-time models of exchange rates and intervention, in: F. van der Ploeg (ed.), The Handbook of International Economics, Basil Blackwell, Cambridge, 251–298

[11] BLACK, F. (1986), Noise, Journal of Finance, 41, 529–543

[12] BLANCHARD, O. (1979A), Backward and forward solutions for economies with rational expectations, American Economic Review, Papers and Proceedings, 69, 114–118

[13] BLANCHARD, O. (1979B), Speculative bubbles, crashes and rational expectations, Economics Letters, 3, 387–389

[14] BLANCHARD, O. AND C. KAHN (1980), The solution of linear difference models under rational expectations, Econometrica, 48, 1305–11

[15] BLANCHARD, O. AND M. WATSON (1982), Bubbles, rational expectations and financial markets, in: P. Wachtel (ed.), Crisis in the Economic and Financial Structure, Lexington Book, 295–316

[16] BLANCHARD, O. AND S. FISCHER (1989), Lectures on macroeconomics, MIT University Press, Cambridge

[17] BROCK, W. (1974), Money and growth: The case of long run perfect foresight, International Economic Review, 15, 750–777

[18] BROCK, W. (1975), A simple perfect foresight monetary model, Journal of Monetary Economics, 16, 133–150

[19] BROCK, W. (1982), Speculative hyperinflations in maximizing models: Corrigendum and clarification, Working Paper, University of Wisconsin

[20] BROZE, L, C. GOURIEROUX AND A. SZAFARZ (1985), Solutions of dynamic linear rational expectations models, Econometric Theory, 26, 341–368

[21] BROZE, L. AND A. SZAFARZ (1991), The econometric analysis of non-uniqueness in rational expectations models, North Holland, Amsterdam

[22] BRUNS, C. (1994), Bubbles und Excess Volatility auf dem deutschen Aktienmarkt, Deutscher Universitäts Verlag, Wiesbaden

[23] BUITER, W. AND P. PESENTI (1990), Rational speculative bubbles in an exchange rate target zone, NBER Working Paper No. 3467

[24] BURMEISTER, E. AND K. WALL (1982), Kalman filtering estimation of unobserved rational expectations with an application to the German hyperinflation, Journal of Econometrics, 20, 255–284

[25] BURMEISTER, E. AND K. WALL (1987), Unobserved rational expectations and the German hyperinflation with endogenous money supply, International Economic Review, 28, 15–32

[26] CAGAN, P. (1956), The monetary dynamics of hyperinflation, in: M. Friedman (ed.), Studies in the Quantity Theory of Money, University of Chicago Press, Chicago

[27] CAMERER, C. (1989), Bubbles and fads in asset prices, Journal of Economic Surveys, 3, 3–41

[28] CAMPBELL, J. AND R. SHILLER (1987), Cointegration and tests of present value models, Journal of Political Economy, 95, 1062–1088

[29] Campbell, J. and R. Shiller (1988), Stock price earnings and expected dividends, Journal of Finance, 43, 661–676

[30] Campbell, J. and R. Shiller (1989), The dividend–price ratio and expectations of future dividends factors, Review of Financial Studies, 1, 195–228

[31] Casella, A. (1989), Testing for rational bubbles with exogenous or endogenous fundamentals. The German hyperinflation once more, Journal of Monetary Economics, 24, 109 - 122

[32] Cass, D. and K. Shell (1983), Do sunspots matter, Journal of Political Economy, 91, 193–227

[33] Charemza, W. and D. Deadman (1992), Rational and intrinsic bubbles: A reinterpretation of empirical results, Discussion Paper, University of Leicester

[34] Charemza, W. and D. Deadman (1995), Speculative bubbles with stochastic explosive roots: The failure of unit root testing, Journal of Empirical Finance, 2, 153–163

[35] Chow, Y. and H. Teicher (1988), Probability theory; Independence, interchangeability, martingales, 2nd ed., Springer–Verlag, New York

[36] Cochrane, J. (1992), Explaining the variance of price–dividend ratios, The Review of Financial Studies, 5, 243–280

[37] d'Autumne, A. (1990), On the solution of linear difference equations with rational expectations, Review of Economic Studies, 57, 677–688

[38] De Bondt W. and R. Thaler (1985), Does the stock market overreact?, Journal of Finance, 40, 793–805

[39] De Bondt, W. and R. Thaler (1990), Do security analysts overreact?, American Economic Review, Papers and Proceedings, 80, 52–57

[40] De Long, J. and A. Shleifer (1991), The stock market bubble of 1929, Journal of Economic History, 51, 675–700

[41] De Long, J., Shleifer, A., Summer, L. and R. Waldmann (1990a), Noise trader risk in financial markets, Journal of Political Economics, 98, 703–738

[42] De Long, J., Shleifer, A., Summer, L. and R. Waldmann (1990b), Positive feedback investment strategies and destabilizing rational speculation, Journal of Finance, 45, 379–395

[43] Dezhbakhsh, H. and A. Demirguc–Kunt (1990), On the presence of speculative bubbles in stock prices, Journal of Financial and Quantitative Analysis, 25, 101–112

[44] Diba, B. (1987), A critique of variance bounds test for monetary exchange rate models, Journal of Money, Credit and Banking, 19, 104–111

[45] DIBA, B. (1990), Bubbles and stock–price volatility, in: Dwyer, G. and R. Hafer (eds.), The stock market: Bubbles, volatility, and Chaos, Economic Policy Conference of the Federal Reserve Bank of St. Louis, Kluwer Academic Publishers, Boston, 9–26

[46] DIBA, B. AND H. GROSSMAN (1985), The impossibility of rational bubbles, NBER Working Paper No. 1615

[47] DIBA, B. AND H. GROSSMAN (1988A), Explosive rational bubbles in stock prices?, American Economic Review, 78, 520–530

[48] DIBA, B. AND H. GROSSMAN (1988B), Rational inflationary bubbles, Journal of Monetary Economics, 21, 35–46

[49] DIBA, B. AND H. GROSSMAN (1988C), The theory of rational bubbles in stock prices, Economic Journal, 98, 746–54

[50] DICKEY, D. AND W. FULLER (1979), Distribution of the estimators for autoregressive time series with a unit root, Journal of the American Statistical Association, 74, 427–431

[51] DICKEY, D. AND W. FULLER (1981), Likelihood ratio statistics for autoregressive time series with a unit root, Econometrica, 49, 1057–1072

[52] DOLADO, J., JENKINSON, T. AND S. SOSVILLA–RIVERO (1990), Cointegration and unit roots, Journal of Economic Surveys, 4, 249–273

[53] DOOB, J. (1953), Stochastic processes, John Wiley & Sons, New York

[54] DORNBUSCH, R. (1976), Expectations and exchange rate dynamics, Journal of Political Economy, 84, 1161–76

[55] DUMAS, B. (1994), Partial equilibrium versus general equilibrium models of the international capital market, in: F. van der Ploeg (ed.), The Handbook of International Economics, Basil Blackwell, Cambridge, 299–347

[56] DWYER, G. AND R. HAFER (1990), Do fundamentals, bubbles, or neither explain stock prices? Some international evidence, in: Dwyer, G. and R. Hafer (eds.), The stock market: Bubbles, volatility, and Chaos, Economic Policy Conference of the Federal Reserve Bank of St. Louis, Kluwer Academic Publishers, Boston, 31–68

[57] ENGLE, R. AND B. S. YOO (1987), Forecasting and testing in co–integrated systems, Journal of Econometrics, 35, 143–159

[58] ENGLE, R. AND C. GRANGER (1987), Cointegration and error correction: Representation, estimation and testing, Econometrica, 55, 251–276

[59] ENGSTED, T. (1993A), Present value models and non–stationary time–series; an introduction, and a summery of the thesis, Institut for Informationsbehandling, Aarhus

[60] ENGSTED, T. (1993B), Testing for rational inflationary bubbles: the case of Argentina, Brazil and Israel, Applied Economics, 25, 667–674

[61] EVANS, G. (1986), A test for speculative bubbles in the sterling–dollar exchange rate: 1981–84, American Economic Review, 76, 621–636

[62] EVANS, G. (1987), The structure of ARMA solutions to a general linear model with rational expectations, Journal of Economic Dynamics and Control, 11, 79–91

[63] EVANS, G. (1991), Pitfalls in testing for explosive bubbles in asset prices, American Economic Review, 81, 922–930

[64] EVANS, G. AND S. HONKAPOHJA (1986), A complete characterization of ARMA–solution to linear rational expectations models, Review of Economic Studies, 53, 227–239

[65] FAMA, E. (1970), Efficient capital markets: A review of theory and empirical work, Journal of Finance, 25, 383–417

[66] FAMA, E. (1976), Foundation of Finance, Basic Books Inc., New York

[67] FAMA, E. (1991), Efficient capital markets II, Journal of Finance, 46, 1575–1617

[68] FLAVIN, M. (1983), Excess volatility in the financial markets: A reassessment of the empirical evidence, Journal of Political Economy, 91, 929–956

[69] FLOOD, R. AND P. GARBER (1980), Market fundamentals versus price–level bubbles, The first tests, Journal of Political Economy, 88, 745–770

[70] FLOOD, R. AND P. GARBER (1983), A model of stochastic process switching, Econometrica, 51, 537–551

[71] FLOOD, R. AND P. GARBER (1984), Collapsing exchange rate regimes: some linear examples, Journal of International Economics, 17, 1–14

[72] FLOOD, R. AND P. GARBER (1991), The linkage between speculative attack and target zone models of exchange rates, Quarterly Journal of Economics, 106, 1367–1372

[73] FLOOD, R. AND R. HODRICK (1986), Asset price volatility, bubbles, and process switching, Journal of Finance, 41, 831–42

[74] FLOOD, R., P. GARBER AND L. SCOTT (1984), Multi–country tests for price level bubbles, Journal of Economic Dynamics and Control, 8, 329–40

[75] FLOOD, R., R. HODRICK AND P. KAPLAN (1986), An evaluation of recent evidence on stock market bubbles, NBER Working Paper No. 1971

[76] FLOOD,R. AND R. HODRICK (1990), On testing for speculative bubbles, Journal of Economic Perspective, 4, 85–101

[77] FRANKEL, J. AND K. FROOT (1986), The dollar as a spekulative bubble, NBER Working Paper No. 1854

[78] FROOT, K. AND M. OBSTFELD (1991A), Exchange–rate dynamics under stochastic regime shifts: A unified approach, Journal of International Economics, 31, 203–229

[79] FROOT, K. AND M. OBSTFELD (1991B), Intrinsic bubbles: The case of stock prices, American Economic Review, 81, 1189–1214

[80] FROOT, K. AND M. OBSTFELD (1991C), Stochastic process switching: Some simple solutions, Econometrica, 59, 241–50

[81] FULLER, W. (1976), Introduction to statistical time series, Wiley & Sons, New York

[82] FUTIA, C. (1981), Rational expectations in stationary linear models, Econometrica, 49, 171–192

[83] GAAB, W. (1990), Möglichkeiten und Grenzen des Erkennens spekulativer Blasen an den Devisenmärkten, Kooperation, Autonomie und Devisenmarkt, in: Filc, W. and Köhler, K. (eds.), Kooperatoin, Autonomie und Devisenmarkt, Duncker und Humblot, Berlin, 67–97

[84] GARBER, P. (1990A), Famous first bubbles, Journal of Economic Perspective, 4, 35–54

[85] GARBER, P. (1990B), The dollar as a bubble, in: Gerlach, S. and P. Petri (eds.), The Economics of the Dollar Cycle, MIT University Press, Cambridge, 129–147

[86] GÖPPL, H. AND H. SCHÜTZ (1992), Die Konzeption eines Deutschen Aktienindexes für Forschungszwecke (DAFOX), Diskussionspapier des Instituts für Entscheidungstheorie und Unternehmensforschung, Universität Karlruhe

[87] GOURIEROUX, C., J. LAFFONT AND A. MONFORT (1982), Rational expectations in dynamic linear models: analysis of the solutions, Econometrica, 50, 409–425

[88] GRANGER, C. AND N. SWANSON (1994), An introduction to stochastic unit root processes, Discussion Paper, University of California, San Diego

[89] GRAY, A. (1984), Dynamic instability in rational expectation models, International Economic Review, 25, 93–122

[90] GROSSMAN, S. AND J. STIGLITZ (1980), On the impossibility of informationally efficient markets, American Economic Review, 70, 393–408

[91] GROSSMAN, S. AND R. SHILLER (1981), The determinants of the variability of stock market prices, American Economic Review, 71, 222–227

[92] GUESNERIE, R. (1986), Stationary sunspot equilibria in an n commodity world, Journal of Economic Theory, 40, 103–127

[93] HALDRUP, N. (1990), Tests for unit roots with a maintained trend when the true data generating process is a random walk with drift, Memo, Aarhus University

[94] HALDRUP, N. AND S. HYLLEBERG (1989), Unit roots and deterministic trends, with yet another comment on the existence and interpretation of a unit root in U.S. GNP, Memo, Aarhus University

[95] HALDRUP, N. AND S. HYLLEBERG (1995), A note on the distribution of the least squares estimator of a random walk with drift: Some analytical evidence, Economics Letters, 48, 221–228

[96] HALL, R. (1978), Stochastic implications of the life–cycle–permanent income hypothesis: Theory and evidence, Journal of Political Economy, 86, 971–88

[97] HAMILTON, J. (1986), On testing for self–fulfilling speculative price bubbles, International Economic Review, 27, 545–552

[98] HAMILTON, J. AND C. WHITEMAN (1985), The observable implications of self–fulfilling expectations, Journal of Monetary Economics, 16, 353–373

[99] HANSEN, G. (1993), Quantitative Wirtschaftsforschung, Verlag Vahlen, München

[100] HANSEN, L. AND T. SARGENT (1980), Formulating and estimating dynamic linear rational expectations models, Journal of Economic Dynamics and Control, 2, 7–46

[101] HERI, E. (1986), Irrationales rational gesehen: Eine Übersicht über die Theorie der bubbles, Schweizerische Zeitschrift für Volkswirtschaft und Statistik, 163–185

[102] HODRICK, R. (1987), The empirical evidence on the efficiency of forward and future foreign exchange markets, Harwood Academic Publishers, Chur

[103] HODRICK, R. (1990), Volatility in the foreign exchange and stock markets: Is it excessive?, American Economic Review, Papers and Proceedings,80, 186–191

[104] HUANG, R. (1981), Efficiency and the volatility of floating exchange rates, Journal of Finance, 36, 31–41

[105] HYLLEBERG, S. AND G. MIZON (1989), A note on the distribution of the least squares estimator of a random walk with drift, Economics Letters, 29, 225–230

[106] IKEDA, S. AND A. SHIBATA (1992), Fundamentals–dependent bubbles in stock prices, Journal of Monetary Economics, 30, 143–168

[107] JOHANSEN, S. (1988), Statistical analysis of cointegration vectors, Journal of Economic Dynamics and Control, 12, 231–254

[108] JOHANSEN, S. (1992), Cointegration in partial systems and the efficiency of single equation analysis, Journal o f Econometrics, 52, 389–402

[109] KAHNEMAN, D. AND A. TVERSKY (1982), Intuitive prediction: Biases and corrective procedures, in: Kahneman, D., Slovic, P. and A. Tversky (eds.), Judgement under uncertainty: Heuristics and biases, Cambridge University Press, London

[110] KALYVITIS, S. AND N. PITTIS (1994), Testing for exchange rate bubbles using variance inequalities, Journal of Macroeconomics, 16, 359–367

[111] KARLIN, S. AND H. TAYLOR (1976), A first course in stochastic processes, Academic Press, New York

[112] KEARNEY, C. AND R. MACDONALD (1990), Rational expectations, bubbles and monetary models of the exchange Rate: The Australian/US Dollor rate during the recent float, Australian Economic Papers, 1–20

[113] KEYNES, J. (1936), The general theory of employment, interest and money, Harcourt, Brace and World, New York

[114] KINDLEBERGER, C. (1978), Manias, panics, and crashes, Basic Books Inc., New York

[115] KINGSTON, G. (1982), The semi–log portfolio balance schedule is tenuous, Journal of Monetary Economics, 389–399

[116] KLEIDON, A. (1986A), Bias in small sample tests of stock price rationality, Journal of Business, 59, 237–261

[117] KLEIDON, A. (1986B), Variance bounds tests and stock price valuation models, Journal of Political Economy, 94, 953–1001

[118] KLEIDON, A. (1988), Bubbles, fads and stock price volatility tests: A partial evaluation: Discussion, Journal of Finance, 43, 656–659

[119] KÖDDERMANN, R. (1993), Chartisten, Bubbles und Handelsvolumen; drei Essays zur Kapitalmarkttheorie, Diss. 1993

[120] KOHN, W. (1991), Eine ökonometrische Analyse von Wechselkursmodellen unter Berücksichtigung nichtstationärer Zeitreihen, Schriften zur angewandten Ökonometrie, Heft 22, Haag+Herchen Verlag, Frankfurt/Main

[121] KRUGMAN, P. (1991), Target zones and exchange rate dynamics, Quarterly Journal of Economics, 106, 669–682

[122] LEHMANN, B. (1990), Fads, martingales, and market efficiency, Quarterly Journal of Economics, 40, 1–28

[123] LeRoy, S. (1982), Expectations models of asset prices: A survey of theory, Journal of Finance, 37, 185–217

[124] LeRoy, S. (1984), Efficiency and the variability of asset prices, American Economic Review, 74, 183–187

[125] LeRoy, S. (1989), Efficient capital markets and martingales, Journal of Economic Literature, 27, 1583–1621

[126] LeRoy, S. and C. LaCivita (1981), Risk aversion and the dispersion of asset prices, Journal of Business, 54, 535–547

[127] LeRoy, S. and R. Porter (1981), The present value relation: Test based on variance bounds, Econometrica, 49, 555–574

[128] LeRoy, S. and W. Parke (1992), Stock price volatility: Tests based on the geometric random walk, American Economic Review, 82, 981–992

[129] Levich, R. (1985), Empirical studies of exchange rates: price behavior, rate determination and market efficiency; in: R. Jones and P. Kenen (eds.), Handbook of International Economics, Vol. II, North Holland, 979–1040

[130] Levich, R. (1989), Is the foreign exchange market efficient, Oxford Review of Economic Policy, 5, 40–60

[131] Lucas, R. (1973), Some international evidence on output–inflation tradeoffs, American Economic Review, 63, 326–334

[132] Lucas, R. (1982), Interest rates and currency prices in a two–country world, Journal of Monetary Economics, 10, 335–360

[133] Lux, T. (1995), Herd behaviour, bubbles and crashes, The Economic Journal, 105, 881–896

[134] Malliaris, A. and W. Brock (1982), Stochastic methods in economics and finance, North Holland, Amsterdam

[135] Mankiw, G., D. Romer and M. Shapiro (1985), An unbiased reexamination of stock market volatility, Journal of Finance, 40, 677–687

[136] Marsh, T. and R. Merton (1986), Dividend variability and variance bounds tests for the rationality of stock market prices, American Economic Review, 76, 483–498

[137] Mattey, J. and R. Meese (1986), Empirical assessment of present value relations, Econometric Reviews, 5, 171–234

[138] McCallum, B. (1976), Rational expectations and the natural rate hypothesis: Some consistent estimates, Econometrica, 44, 43–52

[139] McCallum, B. (1983), On non–uniqueness in rational expectations models: An attempt at perspective, Journal of Monetary Economics, 11, 139–68

[140] MEESE, R. (1986), Testing for bubbles in exchange markets: A case for sparkling rate? Journal of Political Economy 94, 345–373

[141] MEESE, R. AND K. SINGLETON (1983), Rational expectations, and the volatility of floating exchange rates, International Economic Review, 24, 721–733

[142] MEYER, B. (1994), Der Overreaction–Effekt am deutschen Aktienmarkt, Schriftenreihe dr SGZ–Bank, Band 8, Fritz KnapVerlag , Frankfurt/Main

[143] MICHENER, R. (1982), Variance bounds in a simple model of asset pricing, Journal of Political Economy, 90, 166–175

[144] MILLER, M. AND F. MODIGLIANI (1961), Dividend policy, growth and the valuation of shares, Journal of Business, 34, 411–433

[145] MILLER, M. AND P. WELLER (1990), Currency bubbles which affect fundamentals: a qualitative treatment, The Economic Journal, 100, No 400 Supplement, 170–179

[146] MOOD, A., F. GRAYBILL AND D. BOES (1974), Introduction to the theory of statistics, McGraw–Hill, London

[147] MUSSA, M. (1976), The exchange rate, the balance of payments, and monetary policy under a regime of controlled floating, Scandinavian Journal of Economics, 78, 229–248

[148] MUTH, J. (1961), Rational expectations and the theory of price movements, Econometrica, 29, 315–335

[149] NELSON, C. AND C. PLOSSNER (1982), Trends and random walks in macroeconomic time series, Journal of Monetary Economics, 10, 139–162

[150] NEWEY, W. AND K. WEST (1987), A simple. positive semi–definite, heteroskedasticity and autocorrelation consistent covariance matrix, Econometrica, 55, 703–708

[151] O'CONNELL, S. AND S. ZELDES (1988), Rational Ponzi Games, International Economic Review, 29, 431–450

[152] OBSTFELD, M. (1987), Peso problems, bubbles and risk in the empirical assessment of exchange rate behavior, NBER Working Paper No. 2203

[153] OBSTFELD, M. AND K. ROGOFF (1983), Speculative hyperinflations in maximizing models: can we rule them out?, Journal of Political Economy, 91, 675–687

[154] OBSTFELD, M. AND K. ROGOFF (1986), Ruling out divergent speculative bubbles, Journal of Monetary Economics, 17, 349–362

[155] PELLONI, A. (1993) Long–Run Consequences of Finite Exchange Rate bubbles, Open Economies Review, 4, 5–26.

[156] PHILLIPS, P. (1987), Time series regression with a unit root, Econometrica, 55, 277–301

[157] PHILLIPS, P. AND P. PERRON (1988), Testing for a unit root in time series, Biometrica, 75, 335–346

[158] PINDYCK, R. (1984), Risk, inflation and the stock market, American Economic Review, 74, 335–51

[159] PITTIS, N. (1993), On the exchange rate of the Dollar: market fundamentals versus speculative bubbles, The Manchester School, 61, 167–184

[160] POTERBA, J. AND L. SUMMERS (1986), The persistence and volatility of stock market fluctuations, American Economic Review, 76, 435–458

[161] RAPPOPORT, P. AND E. WHITE (1994), The New York stock market in the 1920s and 1930s: did stock prices move together too much? NBER Working Paper No. 4627

[162] RESNIK, S. (1992), Adventures in stochastic processes, Birkhäuser, Bosten

[163] SAMUELSON, P. (1965), Proof that properly anticipated prices fluctuate randomly, Industrial Management Review, 6, 41–49

[164] SAMUELSON, P. (1973), Proof that properly discounted present values of assets vibrate randomly, Bell Journal of Economics and Management Science, 4, 369–374

[165] SCHWERT, W. (1990A), Stock market volatility, Financial Analysts Journal, 46, 23–34

[166] SCHWERT, W. (1990B), Stock volatility and the crash of '87, The Review of Financial Studies, 3, 77–102

[167] SHEFFRIN, R. (1983), Rational expectations, MIT University Press, Cambridge

[168] SHILLER, R. (1979), The volatility of long–term interest rates and expectations models of the term structure; Journal of Political Economy, 87, 1190–1219

[169] SHILLER, R. (1981B), Do stock prices move too much to be justified by subsequent changes in dividends? American Economic Review, 71, 421–436

[170] SHILLER, R. (1981C), The use of volatility measures in assessing market efficiency, Journal of Finance, 36, 291–304

[171] SHILLER, R. (1984), Stock prices and social dynamics, Brookings Papers on Economic Activity, 2, 457–510

[172] SHILLER, R. (1985), Discussion (of Mankiw et al.), Journal of Finance, 40, 688–689

[173] SHILLER, R. (1986A), Comments on Miller and on Kleidon, Journal of Business, 59, S501–S505

[174] Shiller, R. (1986b), The Marsh–Merton model of managers' smoothing of dividends, American Economic Review, 76, 499–503

[175] Shiller, R. (1988a), Fashions, fads, and bubbles in financial markets, in: J. Coffee, S. Ackerman and L. Lowenstein (eds.), Knights, raiders and targets, Oxford University Press, Oxford, 56–68

[176] Shiller, R. (1988b), The probability of gross violations of a present value variance inequality, Journal of Political Economy, 96, 1089–1092

[177] Shiller, R. (1988c), Stock market volatility: An introductory survey, in: R. Shiller, Market volatility, MIT University Press, Cambridge, 77–104

[178] Shiller, R. (1989), Investor behavior in the october 1987 stock market crash: survey evidence, in: R. Shiller, Market volatility, MIT University Press, Cambridge, 379–402

[179] Shiller, R. (1990), Speculative prices and popular models, Journal of Economic Perspective, 4, 55–65

[180] Shiller, R. (1992), Who's minding the store?, in: The Report of the Twentieth Century Fund Task Force on Market Speculation and Corporate Governance, New York

[181] Shiller, R. and P. Perron (1985), Testing the random walk hypothesis: Power versus frequency of observation, Economics Letters, 18, 381–386

[182] Shiryaev, A. (1989), Probability, 2nd ed., Springer–Verlag, New York

[183] Shleifer, A. and L. Summers (1990), The noise trader approach to finance, Journal of Economic Perspective, 4, 19–33

[184] Singleton, K. (1979), On the estimation of linear macroeconomic models with rational expectations, Working Paper, University of Virginia

[185] Singleton, K. (1980), Expectations models of the term structure and implied variance bounds, Journal of Political Economy, 88, 1159–1176

[186] Smith, R. (1990), Self–fulfilling expectations in exchange rates, Economics discussion papers 9004, University of Otago

[187] Spanos, A. (1986), Statistical foundations of econometric modelling, Cambridge University Press, Cambridge

[188] Spear, S. (1984), Sufficient conditions for the existence of sunspot equilibria, Journal of Economic Theory, 34, 360–370

[189] Summers, L. (1986), Does the stock market rationally reflect fundamental values?, Journal of Finance, 41, 591–601

[190] Taylor, H. (1990), Martingales and random walks, in: Heyman, D. and M. Sobel (eds.), Stochastic models, Handbooks in Operations Research and Management Science, Vol. 2, North Holland, Amsterdam

[191] TAYLOR, J. (1977), Conditions for unique solutions in stochastic macro- economic models with rational expectations, Econometrica, 45, 1377–1386

[192] TIROLE, J. (1982), On the possibility of speculation under rational expectations, Econometrica, 50, 1163–1181

[193] TIROLE, J. (1985), Asset bubbles and overlapping generations, Econometrica, 53, 1499–1528

[194] TIROLE, J. (1990), Intertemporal efficiency, intergenerational transfers, and asset pricing: An introduction, in: Champsaur, P. et al. (eds.), Essays in Honor of Edmond Malinvaud, Vol 1: Microeconomics, MIT University Press, Cambridge, 116–144

[195] TOPOL, R. (1991), Bubbles and volatility of stock prices: Effect of mimetic contagion, The Economic Journal, 101, 786–800

[196] WEIL, P. (1987), Confidence and the real value of money in an overlapping generation model, Quarterly Journal of Economics, 102, 1–22

[197] WEIL, P. (1989), Overlapping families of infinitely–lived agents, Journal of Public Economics, 38, 183–198

[198] WEST, K. (1986), A standard monetary model and the variability of the DeutschMark–Dollar exchange rate, NBER Discussion Paper No. 2102

[199] WEST, K. (1987), A specification test for speculative bubbles, Quarterly Journal of Economics, 102, 553–579

[200] WEST, K. (1988A), Bubbles, fads and stock price volatility test, A partial evaluation, Journal of Finance, 43, 639–660

[201] WEST, K. (1988B), Dividend innovations and stock price volatility, Econometrica, 56, 37–61

[202] WEST, K. (1989), Estimation of linear rational expectations models in the presence of deterministic terms, Journal of Monetary Economics, 24, 437–42

[203] WHITE, E. (1990), The stock market boom and crash of 1929 revisited, Journal of Economic Perspective, 4, 67–83

[204] WILLIAMSON, J. AND M. MILLER (1987), Targets and indicators: A blueprint for the international coordination of economic policy, Institute for International Economics, Washington DC

[205] WORKING, H. (1934), A random difference series for use in the analysis of time series, Journal of the American Statistic Association, 29, 11–24

List of Figures

List of Tables

GPSR Compliance

The European Union's (EU) General Product Safety Regulation (GPSR) is a set of rules that requires consumer products to be safe and our obligations to ensure this.

If you have any concerns about our products, you can contact us on ProductSafety@springernature.com

In case Publisher is established outside the EU, the EU authorized representative is:

Springer Nature Customer Service Center GmbH
Europaplatz 3
69115 Heidelberg, Germany

Batch number: 10844669

Printed by Printforce, the Netherlands